Intelligent robotics

Open University Press Robotics Series

Edited by

P.G. Davey CBE MA MIEE MBCS C.Eng

This series is designed to give undergraduate, graduate and practising engineers access to this fast developing field and provide an understanding of the essentials both of robot design and of the implementation of complete robot systems for CIM and FMS. Individual titles are oriented either towards industrial practice and current experience or towards those areas where research is actively advancing to bring new robot systems and capabilities into production.

The design and overall editorship of the series are due to Peter Davey, Managing Director of Meta Machines Limited, Abingdon; Fellow of St Cross College, Oxford University; and formerly Co-ordinator of the UK Science and Engineering Research Council's Programme in Industrial Robotics.

His wide ranging responsibilities and international involvement in robotics research and development endow the series with unusual quality and authority.

TITLES IN THE SERIES

Industrial Robot Applications	E. Appleton and D.J. Williams
Robotics: An Introduction	D. McCloy and M. Harris
Intelligent Robotics	M.H. Lee
Printed Circuit Board Assembly	P.J.W. Noble
Robots in Assembly	A. Redford and E. Lo
Robot Sensors and Transducers	R. Ruocco

ANDERSONIAN LIBRARY

1 2 JUN 1997

BOUND

Intelligent robotics

Mark H. Lee

HALSTED PRESS
John Wiley & Sons
New York — Toronto
and
OPEN UNIVERSITY PRESS
Milton Keynes

To all those who have been neglected

Open University Press
12 Cofferidge Close
Stony Stratford
Milton Keynes MK11 1BY

First Published 1989

Copyright © Mark H. Lee 1989

All rights reserved. No part of this publication may be
reproduced, stored in a retrieval system or transmitted in
any form or by any means, without written permission from the
publisher.

British Library Cataloguing in Publication Data

Lee, Mark H.
Intelligent Robotics.
1. Robots. Development of applications of
artificial intelligence
I. Title
629.8'92

ISBN 0-335-15421-2
ISBN 0-335-15420-4 pbk

Published in the USA, Canada and Latin America by
Halsted Press, a division of John Wiley & Sons, Inc., New York

Library of Congress Cataloging-in-Publication Data
Lee, Mark H.
Intelligent Robotics.

(Open University Press robotics series)
1. Robotics. 2. Artificial intelligence. I. Title.
II. Series
TJ211.L44 1989 629'8'92 88-30417
ISBN 0-470-21393-0

Printed in Great Britain

Contents

Series editor's preface ix
Preface xi

Chapter 1 Setting the scene 1
  1.1   Aiming for realistic goals 1
  1.2   Artificial intelligence 2
  1.3   Industrial emphasis 4
  1.4   Robots: what are they? 5
  1.5   Sensing, thinking and acting 7
  1.6   A future scenario 8
  1.7   Summary 10
  1.8   Further reading material 12

Chapter 2 Sensing the world 13
  2.1   Sensor selection 13
  2.2   Sensor deployment and integration 20
  2.3   Control and coordination 22
  2.4   The importance of constraints 24
  2.5   Summary 25
  2.6   Further reading material 26

Chapter 3 Artificial sight 27
  3.1   A salient characteristic 27

3.2	Computer vision applications	28
3.3	Industrial requirements	29
3.4	Research fields	29
3.5	Pattern recognition	31
3.6	Basic techniques — local operators and segmentation	35
3.7	Industrial vision systems	45
3.8	Future developments	50
3.9	Summary	51
3.10	Further reading material	52

Chapter 4 The problem of perception — 53

4.1	Perception involves interpretation	53
4.2	The analysis of three-dimensional scenes	58
4.3	Blocks worlds	61
4.4	Current research directions	63
4.5	The problem of understanding	65
4.6	Summary	65
4.7	Further reading material	65

Chapter 5 Building a knowledge base — 67

5.1	Introduction	67
5.2	Knowledge representation schemes	69
5.3	Representation review	89
5.4	Knowledge integrity properties	91
5.5	Organization and control	95
5.6	Summary	96
5.7	Further reading material	97

Chapter 6 Machinery for thinking about actions — 98

6.1	Introduction	98
6.2	Searching for solutions	99
6.3	Goal directed planning	118
6.4	Rule based planning	123
6.5	Blackboard systems	126
6.6	Summary	128
6.7	Further reading material	129

Chapter 7 Speech and language: from mouse to man — 131

7.1	The nature of the problem	132
7.2	Speech processing	135

7.3	Text and language analysis	141
7.4	Robotics and factory systems	142
7.5	Summary	147
7.6	Further reading material	147

Chapter 8 Emulating the expert — 148
8.1	The basic expert	148
8.2	A few difficulties and the need for research	152
8.3	Expert systems in industrial automation	155
8.4	The generate-and-test approach	162
8.5	Towards deeper levels of understanding	164
8.6	New techniques for mechanical systems	169
8.7	Summary	174
8.8	Further reading material	174

Chapter 9 Errors, failures and disasters — 175
9.1	The importance of automatic error diagnosis and recovery	175
9.2	Classes of errors	176
9.3	Observed behaviour and internal states	177
9.4	Failures in the assembly world	179
9.5	Coping with errors	182
9.6	Building a world model	190
9.7	Summary	194

Chapter 10 Better by design — 195
10.1	A proposed assembly system	195
10.2	Summary	199
10.2	Further reading material	200

Chapter 11 Towards a science of physical manipulation — 201
11.1	Introduction	201
11.2	The mathematical world	202
11.3	The industrial world	202
11.4	The human world	203
11.5	A case study	204
11.6	Feasibility and maturity	206

Index — 208

Series editor's preface

An industrial robot routinely carrying out an assembly or welding task is an impressive sight. More important, when operated within its design conditions it is a reliable production machine which — depending on the manufacturing process being automated — is relatively quick to bring into operation and can often repay its capital cost within a year or two. Yet first impressions can be deceptive: if the workpieces deviate somewhat in size or position, or, worse, if a gripper slips or a feeder jams the whole system may halt and look very unimpressive indeed. This is mainly because the sum total of the system's *knowledge* is simply a list of a few variables describing a sequence of positions in space; the means of moving from one to the next; how to react to a few input signals; and how to give a few output commands to associated machines.

The acquisition, orderly retention and effective use of *knowledge* are the crucial missing techniques whose inclusion over the coming years will transform today's industrial robot into a truly robotic system embodying the 'intelligent connection of perception to action'. The use of computers to implement these techniques is the domain of Artificial Intelligence (AI) (machine intelligence). Evidently, it is an essential ingredient in the future development of robotics; yet the relationship between AI practitioners and robotics engineers has been an uneasy one ever since the two disciplines were born.

To make his models at all tractable, the AI scientist has so far had to use so many simplifying assumptions that the industrial robot engineer tends to scorn such work as naïve or irrelevant. On the other hand, to make a cost-effective production system the engineer is often forced to tailor it so specifically to the detailed requirements of his production process that the scientist tends to ignore it as totally *ad-hoc*, unworthy of his attention because seemingly showing no generic attributes whatever.

This book occupies a key position in the architecture of our Open University series because the author is an engineer whose own research field — applying knowledge-based systems to industrial automation — lies exactly in between these two extreme viewpoints.

He shows how essential it is for the viability of AI methods in real-world situations that constraints be applied upon searching, perception, and reasoning by using the maximum possible knowledge about the process being automated; yet also how vital it is for the engineer to ensure that he uses computing techniques which are sufficiently generic to have a 'transfer value' into different robot applications, as well as into harder situations that may unexpectedly crop up in the same task.

The author describes important concepts of sensor processing, perception, knowledge bases, planning and expert systems at a non-trivial level — yet with the use of a minimum of jargon — all of these being seen as crucial to future robot systems in manufacture. Other major topics in AI such as the understanding and synthesis of natural speech are carefully described even though they are not yet seen as important for robotics.

The book uses examples from industrial-robot tasks, but also shows clearly how the same AI techniques will benefit the entire process of Computer Integrated Manufacture extending from product design through process planning to automatic manufacture, and beyond that to the harder problems in applying so-called 'advanced' and mobile robots to less structured environments outside the factory.

It will, I hope, come to be seen both as good AI and good robotics, and so do much to foster the mutual support that each discipline can give the other.

P. G. Davey

Preface

This book is an attempt to present the problems and issues that should to be considered when intelligent factory automation systems are being designed. In particular, it is about advanced software for the control of manufacturing systems that create, handle and assemble components. Such systems will act autonomously and aim to optimize their performance in an intelligent, flexible manner.

The author's intention is to provide a considered overview of the nature of artificial intelligence (AI) and its potential applications in industrial robotics. This is not a collection of techniques, it does not contain worked examples and does not describe all the intricate details of the AI programs discussed: this would increase the text tenfold and such compendiums of method are already available. The title contains a pun that reflects the book's central theme: the search for maturity in judgements and decisions when appraising or applying new techniques in advanced automation. The aim is to raise awareness of the nature of next generation software systems that will affect all future industrial activities, especially manufacturing.

The text is addressed to both industrial engineers and computer experts. It is designed for engineers who are familiar with mechanical and electrical systems but would like to know more about the possibilities of flexible intelligent systems and how such facilities might be incorporated into their own working environments. Computer scientists should find the text useful as an exploration of a new and expanding application area from an engineering viewpoint. As well as being useful to automation practitioners, the material could provide the basis for student courses, either as a case study in applied intelligent systems or as part of the software training in an engineering degree. The only assumed background is that readers have some understanding of the basic principles of conventional computer programming, although knowledge of a programming language is not necessary.

It is useful to explain how this treatment differs from other texts on robotics. Although this book addresses engineering issues, it is not about mechanical engineering; it does not cover control systems, kinematics, end-effectors, or other topics which are often featured in robotics texts. For example, regarding sensors, we do not discuss the design and operational details of sensors, nor do we survey the properties of hardware transducers; these can be found in catalogues and in other text books. Instead, we deal with the software problems of organization and control. We examine the functional role of different sensors and the problems associated with processing sensory data. Similarly, we also do not cover robot programming languages because, although intelligent systems will often require sophisticated programming facilities, such languages are a study in their own right and are not a major part of the intelligence issues considered here.

The book deals with the application of knowledge in controlling industrial systems. Applications are defined in terms of tasks — a job consists of a series of sensing and acting tasks to achieve a goal in a given physical domain. Our task domain is drawn from assembly and materials handling rather than mobile robots or other more exotic areas. We aim to deal with the demands of intelligent systems when they are incorporated into a manufacturing framework and their various features and operational properties. We have tried to adopt an engineering attitude towards problem analysis, and emphasize the identification of constraints that are inherent in problems and their use in simplifying solutions. In this sense, we advocate a systems approach which works top-down from the nature of the problem towards a satisfactory solution, given the cost and performance constraints for a given requirement.

Chapter 1 introduces a scenario of a robot in a factory and indicates the various problems and aspects of intelligence that may arise. This gives a context for all that follows. Intelligence must involve sensing and Chapter 2 considers the integration of sensing into the control task and shows how sensors introduce many problems of control and organization. Chapter 3 deals with vision sensors in order to give some feeling of the use of this facility in the industrial arena. Here techniques are examined in more detail than in the other chapters to illustrate the current state of practical commercial systems, and to show their limitations when compared with long term research goals. Chapter 4 explains the difference between sensations and perceptions and outlines central problems for all powerful sensing systems. Chapter 5 reviews the main techniques for encoding and representing knowledge in an intelligent software system. In order to solve problems, a knowledge base must have procedures for access and manipulation, and this topic is covered in Chapter 6. Speech and language are major research areas in AI and Chapter 7 reviews these in the light of the communication requirements of our industrial scenario. Chapter 8 deals with expert systems — that class of AI system that attempts to provide consulting facilities similar to a human expert. Chapter 9 considers the problems of fault analysis, diagnosis and automatic error recovery. Chapter 10 reviews the various issues that have been covered and returns to the initial scenario in order to discuss the application of the techniques in that framework. The influence of AI on the design process is also discussed here. Chapter 11 examines the critical differences between intelligent systems that are capable of flexibly performing industrial tasks and the long term possibilities of systems that emulate human performance.

Each chapter has been provided with a summary which lists the main points that are covered. A short bibliography after each chapter gives key entry points into the literature for following up the techniques that have been mentioned. The selection is chosen to provide further expansion where needed for individual reader's requirements.

Artificial intelligence can be seen as machine thinking, and thinking involves two main aspects: perception and planning. Perception is 'thinking about sensory data and its meaning', while planning is 'thinking about how to perform an action, or series of actions, to achieve a desired goal'. Robotics can be seen as the combination of these two styles of thinking in which sensory data is used to modulate and control actions to achieve a particular goal.

The first area, perception, is a major problem in artificial intelligence, and is not likely to be solved completely for many years, if at all. The problem of deducing the meaning of sensory signals in terms of events in the world does not *seem* very difficult. However, we begin to see why this is an extremely hard problem when we realize that there is usually no unique relation between sensory data and perceptions; wildly different sensations can result in a single perception. In addition, there are many significant factors which influence our perceptions including context, motivation, semantics and various attention controls. In order to illustrate this problem area, Chapters 3 and 4 concentrate on vision in more depth to show the nature of the perception problem. Although we have used vision, it is quite clear that similar difficulties with perception occur in all other sensory modalities.

The other style of thinking concerns the organization of actions in order to change the current state of the world and reach a desired state. Current methods of robot programming concern the writing of code to instruct the robot how to move from one place to another, when to close its jaws, and other such actions. However, future robot programming will involve descriptions of *what* is to be done rather than *how* to do it. This means that the robot will be given a goal specification and will then effectively plan its own solution and execute the plan until it achieves the desired goal. This is a powerful approach to programming and can offer great flexibility for dealing with uncertainties in the world. The idea of producing plans automatically has been seen by AI as a major aspect of intellectual skill that needs to be automated. Although considerable progress has been made in automatic planning, the real life engineering world has some quite different features from many of the toy or artificial worlds that have so far been studied in AI. In some cases this will help us to reach solutions, in others it may mean that current AI research results are not easy to apply.

These two aspects of perception and planning are both complementary and mutually supportive. If our perceptual powers were so strong that we could 'see' everything that was relevant in the world, then planning would not be a very difficult task. We would see immediately all the various possibilities that could be executed. On the other hand, the more planning skills we have, the more capable we are of reasoning about a series of actions and their results, and the less demands we will place on our sensory and perceptual systems. Clearly, a balance is necessary here so that perception cooperates with planning; the planner should not spend days investigating every avenue in the effectively infinite tree of possible decisions, neither should the perceptual system become obsessed with discovering every nuance of the physical world. An important issue in AI is how to achieve such a balance between facilities or processes that could be used to reach a given goal. A message from AI that comes across strongly, is that the combination of different techniques in an integrated manner often achieves much better results than the single-minded application of one technique taken to extremes. Hopefully, we will be able to illustrate this and convince the reader that such a philosophy will maximize dividends for the successful use of intelligent techniques in industrial robotics and other manufacturing situations.

Acknowledgements

Thanks to Chris Price for a critical reading and to Howard Nicholls, Fred Long and several years of students on the AI course at Aberystwyth for contributing ideas and suggestions. Many others who have shaped my views, particularly colleagues in the Robotics Research Group, are equally appreciated. Thanks also to Janet Bambra, Rosemary Law and Ruth Scott for much help with document preparation. I am also grateful for the support of the University of Auckland, New Zealand and the Science and Engineering Research Council.

Permission to use the original of figure 7.7, which is based on a diagram in Computing Surveys (June 1980, p. 219), has been granted by the Association for Computing Machinery and by the authors.

University College of Wales
Aberystwyth

Chapter 1

Setting the scene

Robotics is that field concerned with the intelligent connection of perception to action

Mike Brady

1.1 Aiming for realistic goals

Robotics is one of those high-technology subjects that tend to suffer from an exaggerated or distorted public image. A news item about a line of welding robots in a car factory will often have a commentary that hints at dire consequences for the work-force, the car industry and even the human race. Why do robots provoke this type of response when, for example, the installation of a new type of machine tool, that might have a greater effect on productivity, goes virtually unnoticed? There are at least two reasons for this — superficial simplification and emotional appeal.

Robotics, like politics and economics and unlike quantum theory or electronics, has an immediacy about it that allows everyone to feel they really understand enough to debate the important issues of the day. This direct appeal to our intuitive understanding stems from the very visual and physical nature of robot hardware (that has been such a bonus for the media, with their hyperbolic reporting, often verging on science fiction!). Of course, this ease of understanding is an illusion. Just because we can easily identify with a problem area does not mean that we have a significant theory of the underlying mechanisms. In fact, this lack of a formal theory is what allows us to get away with superficial arguments and unsound reasoning. There have been some spectacular disasters in politics and economics when an intuitive theory has been pursued to its ultimate conclusion! We aim to show that robotics, being a very new field, also needs a variety of rigorous formal theories to be developed before we can say we fully understand it.

The other factor, emotional appeal, arises out of our response to machines that operate in man's image. Despite our knowing the wide differences between man and machine there is always a fascination in watching even the most blind and dumb factory robot performing its mechanical execution of spatial movements. This emotional connection actually extends to include the automation of human tasks of all kinds and, in particular, the relation between computational processes and human thought. Nowadays, computers are less likely to be referred to as 'electronic brains' but the idea is often implied. It is very important to be careful that any analogies with humans are well founded and realistic if we are to draw meaningful conclusions from them.

Despite the entertainment value of the subject with its speculations and diversions we should all be aware that the public image can get out of hand and become quite unrelated to current research progress and realistic applications. While not ignoring or discouraging the popular side of robotics, the robotics scientist or engineer must be able to separate the expectations of the media and the public from his professional assessment of realistic goals and achievements. The aim of this book is to examine techniques that will enable robots to become more intelligent in the future and to consider the main problems and any potential barriers to progress. The intention is to introduce the broad spectrum of issues and ideas concerning intelligent systems and to develop in the reader an appreciation of how such concepts might be implemented in future robot systems. Although different methods are described, these should not be seen as a collection of immediately useful and proven solutions to problems; intelligent robotics is a research field — most of the work concerns understanding the problems rather than giving cookbook solutions. Rather, the goal of the book is to develop a critical awareness of the feasibility of possible solutions and encourage an engineering approach where all available information and constraints are brought to bear in an effective and appropriate response to a problem. This kind of maturity will enable us to sort out the science fiction from the exciting possibilities with real economic potential. As a test of how successful the book has been in achieving this objective, the author will be pleased if the reader understands (i.e. can give a reasoned analysis), for example, why a fully automatic unmanned factory is a much more realistic goal than the development of a universal general-purpose household robot.

The two main aspects of robotics that this book attempts to relate to each other are intelligent systems and industrial applications.

1.2 Artificial intelligence

Artificial Intelligence (AI), or the better but less used term *Machine Intelligence*, is a subject that is even more controversial than robotics and can be interpreted or defined from many different viewpoints. Rather than enter into a long and complex definition of terminology we will use the term AI to mean any machine implementation of a task that would be considered to require intelligence if performed by a human. Examples are chess playing programs, mathematical reasoning programs and resource planning systems. Of course, if we look at these closely, we may find that they operate in a completely different way from humans. Chess can be played, for example, by generating a brute force search through the tree of legally possible moves. But we will avoid lengthy philosophical discussion about the mechanisms of intelligence and the relations between man and machine by simply adopting the criterion of *successful performance of a complex task*. If a task would usually require human intelligence but a computer or robot system can successfully perform the task, then we consider it to employ some form of *equivalent* intelligence. We will largely ignore aspects of style, i.e. how the goal is achieved, but simply measure the degree of achievement. This will not satisfy the cognitive scientist, who

Setting the scene

tries to understand human thought processes by analogy with AI results, but we are manufacturing engineers who have a different job to do!

Notice that this view is very task specific; in a different task role the same system may appear quite unintelligent. Thus, by our definition, we would consider a pocket calculator to embody *some* intelligence about square roots, exponents and logarithms even though it falls far short of the popular image of an intelligent machine. (In order to convince anyone of the non-triviality of a calculator ask them to consider either how to build one or how to teach someone to perform the equivalent range of mathematical functions.) This definition closely corresponds to the use of the adjectives 'smart' and 'dumb' when applied to electronic equipment. A 'smart' system has its own in-built logic or processes and can perform certain tasks on its own, while a 'dumb' system can only do what it is told. Clearly, a pocket calculator is 'smart' at arithmetic. Also, many industrial robots are able to compute complicated mathematical transforms in order to control their spatial movements; they have in-built smart functions. However, the significant deficiency in both calculators and robots, in terms of intelligence, is their lack of *flexibility*. We shall see that flexibility, in various forms, is a key feature of intelligence.

This task-based definition implicitly refers to the intellectual or computational demands imposed on our robots by the given behavioural or operationally specified task. Having passed the buck from intelligence to tasks, the issue now shifts towards deciding *which tasks* are complex enough to require intelligence. But this is so subjective that we leave this to readers to make up their own personal criteria, on the basis that 'we all know intelligence when we see it'. Some popular perceived facets of intelligence that the author has collected through discussions with students are listed in Figure 1.1. These are offered as a stimulus for adventurous readers interested in exploring the philosophy of AI and its relation to cognitive science. Further material on this topic is listed in section 1.8.

It is important to realize that AI is not solely concerned with the search for solutions to problems but is ultimately concerned with discovering the nature of the problem itself.

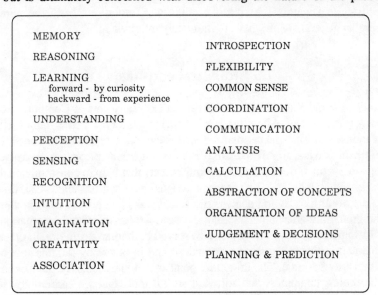

Figure 1.1   Aspects of intelligence

Instead of viewing each AI technique or development as a finished solution (which, in most respects, they rarely are) we can see them as experiments that uncover a little more about the general problem of intelligence and how it can be approached, implemented and utilized. The nature of intelligence is so complex and the field is so vast that simply defining a problem area is a major task. Fortunately, in industrial applications, we are not involved in cognitive modelling and have available much more information about the nature of the problem. Because industrial environments are engineered they are highly structured compared with more 'fuzzy' human problem areas. This means we are able to formulate fairly tight specifications of our automation problems. However, we must always be aware of the potential 'open endedness' of any 'intelligent features'. One of the best ways to expose any vagueness is to write a very detailed specification document. Problem definition is an important stage in any computing project but is vital in AI as it will constrain and illuminate the form of the solution. The more work spent on clarifying, constraining and understanding the problem, the greater is the likelihood of producing an effective solution.

There is a tendency in AI for solutions to become somewhat devalued as they become more widely known. As yesterday's difficult research technique becomes today's commonplace algorithm, so expectations change to new levels. This happens because AI is an intensely active research field constantly throwing out new results across a wide range of topics. Partly because it is a young field, and there is so much to be done, the focus of current research tends to shift quite frequently, giving the impression of a fashion-driven subject. Despite these factors, the core areas in AI are continuously being developed and AI solutions have made many valuable contributions to computing in general. For example, the Macsyma system, developed at MIT in the early 1970s, was a very ambitious symbolic integration and manipulation system that aimed to perform as well as an advanced mathematician. Although AI research still continues on the automation of mathematics, the Macsyma system is now a well-respected software package that is marketed, maintained, and used routinely by scientists and engineers as a tool in their professional work. It would be wrong to try to separate the spin-off value of AI solutions from the pure research programmes that generate them; they are intimately intertwined.

1.3 Industrial emphasis

Although this book is about AI techniques, which are really part of Computer Science, the emphasis is on the use of AI within an industrial context, i.e. the manufacturing and production industries. This means that we will focus on a rather more constrained robot environment than is usual in AI. Industrial robot applications provide a clear specification of a problem area with various pragmatic requirements that help to constrain and, hopefully, simplify the eventual solution. The approach adopted here can be called 'task driven'. This means we first of all look at and analyse the actual tasks that are required in some factory process, and then, only after understanding the task, we consider how automation could be of benefit. In some cases the task will be best served by human operators, in other cases by fixed automatic machinery (i.e. hard automation) and in others by flexible robotic devices (of many and various forms). The important point is that knowledge of the *task* is our main goal and the robots, computers and software are all tools that we can employ to achieve suitable execution of the task. This viewpoint is an attempt to ensure that we do not get

Setting the scene

carried away by the power of some sophisticated technique and modify our interests so that the 'means' become more important than the 'ends'. In computing this effect is well known; programmers often use 'clever' methods in preference to perfectly acceptable simple ones, and managers can't always justify their purchase of fancy computer systems in terms of economic arguments for a well-defined task!

On the other hand, the industrial world is not so constrained that solutions to automation problems are easy to find. Some areas are extremely complex and require advances in the most powerful and sophisticated AI techniques for future progress to be made. Hostile environments provide good examples of such complexity; for example, diving robots for underwater maintenance of oil installations, or space robots for construction and repair work. However, given that a wide spectrum of solution levels may exist, we should try to satisfy the task by the most effective, efficient and simple solution that is available, whether it be considered AI or not.

1.4 Robots: what are they?

There are two important concepts that must be clarified from the outset: what are robot systems and what do they do? Surprisingly, despite everyone's strong intuitive notions about what constitutes a robot, official definitions vary widely in scope. The Japanese definition is so broad that they include practically any physical device operated by some form of program. An example of this category would be the 'robot' cash dispenser in the high street bank. In Britain, the Department of Industry simply specifies a 'reprogrammable manipulator device' while the British Robot Association gives a full and detailed definition. This situation is very unsatisfactory, not least because the estimates for the numbers of robots used by different countries vary enormously, depending upon whose definition is used. We will not enter into the debate on first-generation robots, but instead suggest a simple definition of *intelligent* (second-generation) robots founded on intuitive understanding.

Let us consider an *intelligent robotic device* to be a machine that:

(a) can manipulate physical objects in the real world,
(b) can sense events in the world and
(c) is flexible.

We consider a machine to be flexible if it can *change its task over time, both* by being *reprogrammed* and by *automatic task adaptation* of some form. We don't have to worry about the *shape* of our robot, it might be a manipulator arm, a mobile truck or a distributed machine control system; but it must have sensors and be flexible. No robot can be called intelligent if it is without significant sensing capabilities or has a rigid control sequence. This rules out the cash dispenser mentioned above (even though it might be considered to sense through its keyboard). Likewise non-sensing robots such as paint spraying and welding machines are also unintelligent. Although such first-generation robots are reprogrammable and are a marked improvement over hard automation, future generations of adaptive sensing robots will show a dramatic change in performance as AI techniques are exploited. Therefore, this viewpoint excludes both hard automation (which may incorporate

COMPUTERS	ROBOTS
Input symbols are static and well behaved	Sensory signals are noisy and unreliable
Operations give consistent results	An action can have different responses
Environment is fixed and repeatable	Objects may move about independently
System only receives intended inputs	Influences from external agents can interfere
Perfect performance assumed for computing environment	Operating environment is unreliable, dynamic and incomplete

Figure 1.2 The contrast between environments seen by computers and robots

sensing) and computer controlled machinery (which may be flexible) unless *both* sensing and flexible task performance occur in the same system.

The other issue worth examination is what do robot systems *do*? That is, what is the range and extent of their capabilities? The important difference between a robot and a computer is that the robot performs actions in the real *physical* world while the computer performs operations on symbols in an internal 'information processing' world. This is a consequence of the manipulation and sensing abilities of robots that create different operating regimes from conventional computing. Figure 1.2 illustrates these differences. Robots operate on objects and cause events to occur and transformations to change the state of the world. (By 'world', we mean the local operating environment of the robot and all agents with which it can interact). These events and transformations are often not reversible or repeatable, e.g. when a feeder presents a component or when two parts are glued together. Also, there is frequently a degree of uncertainty as to whether the operation was as successful as intended. Such considerations just do not occur in conventional computing where symbols stay as you left them, computations are repeatable, and checking, restart and monitoring procedures give protection against hardware errors. The symbolic world of the computer program features perfect memory, 100% repeatability of instructions, precise information about the result of each instruction and complete protection from outside interference or external events.

By contrast, in the robot's operating world, as distinct from its computer controller, components do not always stay in the same place; the same movement can sometimes produce different results; there are always some unknowns about aspects of the task; and

Setting the scene 7

external interference is not impossible. This means that designers of robot systems have significant intrinsic difficulties to deal with, *in addition* to those associated with implementing computer-based systems. Of course, we do not suggest that the computing aspects are trivial (far from it, as most of our problems must be solved, in the end, by software techniques), but rather that a shift of viewpoint is necessary. In computing, programmers usually assume a reliable, consistent operating environment and concentrate on designing, implementing and testing their solutions to problems. But in robotics, we have to accept that operating in the real world, with all its physical phenomena and subtle ramifications, we do not have full control over our operating regime. Thus, we must adopt a more defensive style of system design and control. We ought to check each action, monitor the environment frequently and always be prepared for new events. This type of approach will involve much hard work, and certainly demands much more care and consideration than the programming of conventional fixed-cycle robots and industrial machines. However, we shall see that there are valuable benefits to be gained and that, in fact, this approach is essential if unexpected events such as errors or failures are to be safely handled without damage and disruption.

Of course, if we could control and constrain all the equipment and the environment so that extreme reliability was achieved, then programming would be much simpler (and not require AI methods). However, the costs of providing such exceptional quality control and reliability must be weighed against those involved in using techniques to deal with some measure of uncertainty. Also, with the very small batch sizes expected in future flexible factories, it will be much cheaper to provide some form of intelligence that can help with the design, setting-up and operation of manufacturing cells for different yet related products, rather than hand-code each and every aspect of the process. It is likely that frequent design and equipment changes will increase the potential for uncertainty and unexpected events. Also, there will always be some areas that cannot be fully controlled, for example those that involve interaction with human operators.

1.5 Sensing, thinking and acting

We have suggested that the actual shape of the robot system is not significant and also that physical processing rather than symbolic processing needs to be the ultimate focus of our attention. This means we will not be very concerned with machine design and engineering, i.e. the specification of kinematics, speed, performance etc., nor will we be overly interested in feedback control systems, numerical position and spatial controllers or even conventional robot programming languages. As these are all good traditional robotics topics, what then concerns the designer of intelligent robot systems? The difference is that in first-generation robots the task is so well specified and so fixed that the main problem focuses on how to build and/or install a manipulator that can, typically, move objects from A to B under given conditions. When high flexibility is introduced the problem area shifts from the robot design arena to understanding and specifying the task. Let us now make the assumption that objects *can* be moved from A to B (by machines resulting from the efforts of robot design engineers), and deal with task questions such as *which* object should be moved, *why* and to *where*. (Actually, this approach will eventually feed back into robot design, as new features and facilities are requested for various tasks; and so the task-driven approach is really only a re-ordering of priorities.)

Now, by considering the activities that must be performed by an intelligent robot controller able to mastermind flexible task performance, we see three main categories:

1 Sensing involves the acquisition of data about the environment and the robot itself. This is an obvious requirement but the range and nature of the sensors strongly affect the quality of the system's perceptions.
2 'Thinking' means symbolic computation directed at performing some activity, in other words, reasoning about a task. This can include reasoning about both sensory data and actions before, during, or after execution.
3 Acting is not just the initiation of movements but also the planning of movement sequences and their control and coordination.

These three activity areas form the core of our AI philosophy for robotics and are examined in detail in following chapters.

1.6 A future scenario

As an illustration of the above ideas and concepts, a scenario of a futuristic but plausible advanced system can be constructed. This is intended to give the flavour of the term 'intelligence' and to show the kind of features, requirements and performance of future systems that AI techniques will make possible. Consider the following conversation between a human factory supervisor and one of the assembly robots:

SUPERVISOR: How many gearboxes have you assembled?
ROBOT: In the last 24 hours: 27 of type A, 32 of type B, 5 of type D and 8 of type H.
SUPERVISOR: Give reject report on type A.
ROBOT: 26 accepted by tests; of these: 3 had component faults, which were rejected and replaced, 1 had assembly errors, which were rectified. 1 failed tests with undetected errors at assembly stage 21: diagnostic report identifies possible causes.
SUPERVISOR: Explain assembly errors.
ROBOT: Gearbox number 13 failed free rotation test at assembly stage 17. Disassembly to stage 15 gave diagnosis of pin 39a not fully inserted due to gripper slip. Minor slippage detected on 3 previous occasions. Gripper pads diagnosed as worn or excessively greasy. Gripper pads replaced. No further slip detection occurred. Re-assembly to stage 17 and then to completion, passed all tests.
SUPERVISOR: Give performance report on type A.
ROBOT:
Assembly-time computer utilization	52%
Total actuation duty cycle	64%
Sensor activity level	34%
External feed dwell time	7%
Failure processing time	9%

SUPERVISOR: Prepare to request and receive specification of pump type Q from design office.

Setting the scene

ROBOT:	Ready.
SUPERVISOR:	Go ahead.
ROBOT:	Data received from design office computer number 3 and verified.
SUPERVISOR:	Rotor plate BQ7 has been given a temporary modification and is now called plate BQ9 in pump specification. Stand by to examine sample.
ROBOT:	System cleared and ready to examine component. Place in viewing area.
	[*Supervisor places modified part in viewing area, robot scans shape and then grasps part and views from other angles.*]
ROBOT:	Rotor plate BQ9 entered in data base for pump assembly type Q, replaces part BQ7, 5 feature differences noted.
SUPERVISOR:	Prepare to assemble batch of pumps type Q.
ROBOT:	Concurrent with gearbox batches?
SUPERVISOR:	No.
ROBOT:	Gearbox plans dumped. Requested and received pump assembly plan from central line planner. Work cell reconfigured and gripper exchanged. Run-time monitor now loaded with data base, plan and task specific knowledge. Please enter feeder assignments.
SUPERVISOR:	Component trays arrive at feeder point C, assemblies exit from port A, reject assemblies exit from port B and reject components from conveyor D.
ROBOT:	Monitor detects no inconsistencies in trial simulation mode. Response estimates are as follows:

Computer utilization	63%
Actuation duty cycle	70%
Sensor activity	48%
Plan predicts sensory and test integration level as	77%
With potential error detection coverage	93.8%

SUPERVISOR:	Enter assembly mode.
ROBOT:	Assembly mode enabled.

Let us now review the features and capabilities suggested by this conversation. Perhaps the most obvious is the use of speech for communication between human and machine. In fact, for many factory situations this would be unsuitable: a set of buttons and a display might be more appropriate. Also, the dialogue is far more detailed and lengthy than necessary because it is an illustration intended to give a clearer picture of what is happening. Nevertheless, such speech facilities will be possible and could be incorporated into all forms of automated equipment. The other aspect of the communication is its naturalness. When the supervisor instructs 'explain assembly errors' the robot *assumes* the context of the previous exchanges and reports on the *one* assembly failure of type A that was rectified. This sort of 'looseness' occurs all the time in natural language conversations and is an essential feature of human dialogue. However, it is very complex and difficult to handle in a computer and is *not* to be confused with computer programming languages which are, by contrast, highly structured and constrained.

The robot's task seems to consist of a succession of small batches of fairly complex components which it assembles into a complete unit with the help of a given assembly plan. The batches, components and assemblies frequently change and so the likelihood of errors is higher than for long runs of dedicated assembly machinery. The essential flexibility demanded by this work pattern is facilitated by the use of a wide range of sensors. It is clear that the 'tests' referred to must be inspection stages and thus involve a variety of sensory parameters. Likewise, to detect errors during assembly requires good sensory capabilities, as also does the ability to reconfigure the work area.

Such small batch working will become very common in manufacturing. It offers savings in materials, storage and wastage, while giving design flexibility and rapid response to customer demand. Intelligent manufacturing cells will have major advantages for small batch sizes because they will give short set-up times, closely monitored operation and high quality products. As so much production and design data will be computer generated it would be pointless to aim for human programming of the manufacturing cells; another reason why flexibility is so important ....

Regarding the computing resources needed by such a system, these would be considerable, as not only is the robot controller itself very powerful, it is also connected to at least two other computers: the 'central planner' and the design office computer. These must all be of significant size as they process large data files, e.g. the pump and gearbox specifications, the assembly plans and the run-time monitor database. In addition, there are formidable specialist software requirements, as seen in the use of vision to re-define the data-base specification of a given component. Figure 1.3 indicates the computing requirements and the various software modules and packages.

When the supervisor asks a question or gives an instruction, the demands made on the robot range from trivial to highly complex. For example, the first part of the dialogue concerns logging reports. Data logging is a routine technique and presenting the data is simply a matter of file access and processing. On the other hand, generating meaningful diagnostics about why an assembly failed is an extremely complex task. This requires knowledge of the structure of the parts, knowledge about assembly events (e.g. slippage, tool wear), and knowledge about when and how to use sensors to detect the relevant events. In other words, the system must *understand what it is doing*, if we require it to reason about all the possible task variations, and accommodate changes or errors with the least amount of external input.

This scenario will form a basis for discussion as specific areas are examined in more detail in subsequent chapters.

1.7 Summary

1 It is important to be able to estimate what robotics research can reasonably achieve in terms of possibilities, feasibility, and cost effectiveness.
2 Intelligence is a difficult concept that has many definitions and stimulates much controversy and philosophical debate. A useful task-based definition views intelligence as a requirement for the performance of tasks of sufficient intellectual difficulty.
3 It is most important to develop a full understanding of a problem before potential solutions are examined. A written specification will greatly help to clarify the problem requirements. This is a basic computing design principle but it is especially important in AI as the exact nature of problems can be difficult to pin down.
4 Industrial applications should be defined and specified in terms of task goals and processes.

Setting the scene

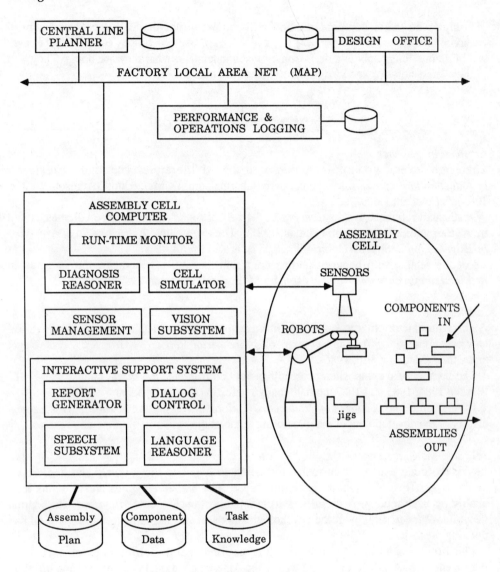

Figure 1.3 Factory scenario

5 The word 'robot' will be used to cover all kinds of industrial machinery provided they can act and sense in the physical world and are flexible in their performance. The shape of a robot is not important.
6 When designing robot control programs for the physical world we must be cautious about external influences and interferences — we should take less for granted than in 'normal' computer programming.
7 Sensing, thinking and acting are the three key areas for progress in intelligent robotics.
8 Future systems will use large amounts of cheap computational power, will run large and

complex software components, and will require sophisticated support and control environments.
9 Future intelligent robots will, in some sense, understand what they are doing. They will have large stores of knowledge about their current tasks.

1.8 Further reading material

Artificial Intelligence
There are several good entry points into the AI literature. The student textbook *Introduction to Artificial Intelligence*, by E. Charniak and D. McDermott (Addison-Wesley, 1985), is excellent value.

A standard reference is *The Handbook of Artificial Intelligence*, in three volumes, edited by A. Barr and E. Feigenbaum (Pitman, 1981). The two volume *Encyclopedia of Artificial Intelligence*, by S. Shapiro (John Wiley, 1987), is also useful.

For the relationship between human cognition and AI, a general review is given in *Artificial Intelligence and Natural Man*, by M. Boden (Harvester, 1987).

Robotics
A good introductory text is *Robotics: An Introduction*, by D. McCloy and M. Harris (Open University Press, 1986). Another textbook is *Industrial Robots* by Groover, Weiss, Nagel and Odrey (McGraw-Hill, 1986).

The international symposium proceedings entitled *Robotics Research*, published by MIT Press in 1984, 1985 and 1987, contain many interesting research papers.

David Noble's book *Forces of Production*, (Oxford University Press, 1986), gives a social history of the development of manufacturing automation.

Journals and Conferences
In AI, the main source of current information is found in the journals and conference proceedings. The chief journal is simply called *Artificial Intelligence* and contains high quality research reports. A useful less formal publication is the *AI Magazine*. The journal *Cognitive Science* deals with the psychological side of AI that aims to understand human thinking processes.

The premier conferences are: the International Joint Conferences on Artificial Intelligence (IJCAI), held every two years; the American national conferences organized by the American Association of Artificial Intelligence (AAAI), held whenever there is no IJCAI on the North American continent; and the European Conferences on Artificial Intelligence (ECAI).

In robotics, the *International Journal of Robotics Research*, (IJRR) is an important research journal, as is the IEEE *Journal of Robotics and Automation*. Many other journals have recently started, e.g. *Robotica, Robotics, Robotics and Computer-Integrated Manufacturing*, and the *International Journal of Robotic Systems*.

Two useful series of conference proceedings are: the *International Conferences on Robot Vision and Sensory Controls* and the *International Symposium on Industrial Robots* (often combined with the International Conference on Industrial Robot Technology).

Chapter 2

Sensing the world

We are astonished at thought, but sensation is equally wonderful

Voltaire

A robot sensor is a transducer that converts physical effects into electrical signals which can be processed by measuring instruments or computers. Just as biological sensors vary in their performance and relative utility so do man-made receptors have different powers and scope. In order to use sensors effectively we must have a sound understanding of sensor technology and its application in advanced automation. It is a truism that sensing is essential for intelligent systems, but the important corollary is that the quality and effectiveness of the sensory system ultimately determines the limits of intelligence. For intelligent robotics it is vital that well-designed and well-organized sensory capabilities are built into the system from the outset.

There are at least three main problems facing the design engineer when sensing is to be incorporated into automated systems. These are the selection, deployment and management of sensors. For a good appreciation of the complexity of these issues it is worth looking at each in turn.

2.1 Sensor selection

If we open a component-manufacturer's catalogue we will find a wide range of sensors of all shapes, sizes and types. How should we choose a sensor for a given robot application? What should be the basis for a decision in selecting one sensor rather than another? Of course, a skilled designer may have enough experience and judgement to be able to find a satisfactory solution very rapidly. But this 'intuitive' approach will not help anyone in the long term. We must be able to justify our design decisions and we must make them explicit and available for the benefit of education, training and further progress. In engineering, analysis and observations drawn from existing systems are used to refine and improve the science of the subject. In this way we move from a collection of 'rules of thumb' towards a theory that represents the current body of knowledge. Unfortunately, AI and robotics have not yet reached this point and so there are no fundamental 'text books' on topics such as

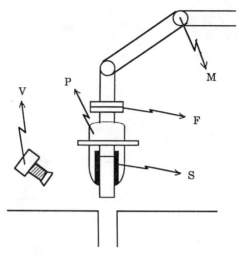

Figure 2.1   Multi-modal sensing

sensor selection. Nevertheless, we must try to consider the options as scientifically as we can.

Consider an example task as a way of illustrating the sensor selection problem. Figure 2.1 shows an insertion process, where a peg is to be inserted a small distance into the hole in a block. Suppose that the peg has plenty of clearance but occasionally (due to undefined errors) the peg is not quite central in the jaws and thus can hit the edge of the hole. Suppose further that we have been asked to design a sensory system that will signal when this error event occurs. The diagram shows five possible sensing methods:

- V vision (for observing the misalignment)
- F wrist force (for detecting stress)
- M motor power (for detecting increase of effort)
- P vertical position (for noticing arm position errors)
- S slip sensing (for feeling relative movement between peg and gripper)

It seems that each of these could satisfy the functional requirement of detecting the specific error, and yet they are entirely different in nearly every respect. In terms of cost, performance, efficiency, generality and even sensed variables, these sensors have widely different parameters. Also, this is only *one* task requirement. If there were many other task activities (as there will be) and each could be solved by five different methods (as they could be) the range of choice becomes very large. Of course, the five methods suggested here are quite arbitrary; you may well think of more! We could even take an extreme view by arguing that because good engineers have a natural ingenuity, *most* sensors can be used in practically *any* application.

However, human designers do not always reason about sensing in such an open, objective manner. They will often reject a particular sensor because it is too costly/unreliable/bulky/etc. By similar reasoning it might be shown that, due to other constraints, there is only one 'obvious' sensor to use. However, the constraints which are so useful in such decisions are not fixed facts of life but change rapidly with technological progress. While a vision sensor might be an expensive system today, it could be a cheap component item tomorrow. So arguments about such practical features of sensors are not really the issue here. There exists a more important theoretical level which will greatly help us in our design work. The theoretical problem is what does (for example) vision offer as compared with other sensors in terms of *function* and *effectiveness* for the task in hand. We would like to know a theoretical basis for our decisions *first*, before applying the practical constraints. Thus, we might determine vision as being the most appropriate or powerful form of sensing for a given task, and yet later adopt a cheaper method due to trade-offs with the practicalities of the latest sensory products. Another factor is that a theoretical approach will help to deal with the complex requirements of sensing for intelligent behaviour. Flexibility is essential in intelligent systems and especially so in their sensory sub-structures. If we simply select sensors on a narrow 'one-sensor for one-problem' basis we will not benefit from the rich interactions between sensors that can be so important. Sensor cooperation and coordination techniques can provide many measurements that are not available to any single sensor alone. These methods also offer increased reliability and more consistent performance. By considering sensors as an integrated facility, rather than as independent signal sources, we gain potential coverage and performance, and thus increase the scope for flexibility and intelligence.

One approach to the sensor selection problem has been to develop classification schemes and place sensors in various kinds of taxonomic structure. Perhaps the most familiar of these are seen in manufacturers' catalogues, where sensors are grouped according to characteristics such as physical sensing mechanisms and operational technical data. Figure 2.2 shows a series of different sensor mechanisms and Figure 2.3 shows some typical operational data. These characteristics are frequently used to help identify desired sensory

---

Mechanical Switching

Resistive effects

Capacitive effects

Inductive effects

Visible light imaging - solid state devices
                           - vacuum tubes

Thermal effects

Photo-electric and infra-red

Ultrasonics

Piezo-electric effects

Hall effect

Figure 2.2   Primary physical mechanisms employed in sensors

Figure 2.3 Typical sensor operational data

needs and can be very useful in eliminating possibil....s. However, there are serious disadvantages with classifications based on such features. For one thing, the operational data are constantly changing, due to incremental refinements and developments. Secondly, the physical mechanisms employed inside the sensor are not, or should not be, of central concern as they detract attention from the objectives of the sensory task. Finally, these classification schemes force the selection process to be one of elimination, rather than design.

Other schemes have been used in the research literature. Some use taxonomies that distinguish sensors by using a set of attributes such as contact/non-contact, scalar/vector or imaging/ranging. Each sensor can then be evaluated in terms of the attribute set. The selection process consists of answering questions about the desired attributes, thus producing a decision path through the tree of sensor types. Figure 2.4 shows a sample classification scheme based on only two attributes. As more attributes are used so the data form a decision tree. These methods are more valuable than the operational type of classification but they are still limited by the original choice of attributes used. The attributes must be decided beforehand and the danger is that they may be influenced by the available sensing technology rather than the more abstract task requirements. After all, as new sensors are invented and developed so new attributes will be added and thus the value of previous decisions based on the taxonomy will change.

Perhaps the best classification method is to focus on the different sensory modalities, that is, not the underlying physics of the sensor but the physical property or event that must be measured. Humans have major modalities such as sight, hearing and touch and lesser modalities such as taste, smell and temperature detection. However, although general

Sensing the world

Attribute 1

Resolution of Output Values

Attribute 2 Output Complexity		Binary	Many Valued
	Scalar	limit switch	temperature probe
	Vector	spatial direction indicator	force sensor
	Array	thresholded imaging	grey scale imaging (vision, thermal, tactile)

Figure 2.4 Examples of sensors classified by two attributes of output signal

analogies with human performance are often stimulating (and are used in this book) it is not very useful to draw too detailed inferences from human sensing, as (a) many of the human senses would have a very limited role outside of the biological setting (e.g. balance control of bipeds), and (b) the range of transducers that technology can provide bears little relation to biological sensors. Not only are many human sensory skills beyond current technology but paradoxically there are devices that can sense in modalities that have no human counterpart. The relations between humans and robotics and the dangers of an anthropomorphic bias are discussed in Chapter 10. If we identify the distinct physical parameters that can be relevant in robotics we find a series of modalities that express our sensing requirements. Figure 2.5 shows a diagram of sensors classified by type, and as a secondary feature, by distance from the robot system. There are three major categories — *internal*, *surface* and *external* sensors — with the external region being divided further according to the distance of the sensed properties. These categories highlight certain fundamental differences. For example, surface and external sensors deal with contact and non-contact events respectively and exclusively, while internal sensors are the proprioceptors — sensors which deal with internal states almost completely independently of surface and external events.

This classification by modality of received sensation is useful in identifying general classes of desirable sensor but it leads on to an even higher level scheme of greater value. In our original peg-in-hole problem we have several candidate methods of detecting a failure. If we postpone the practical arguments (e.g. 'vision is too expensive') and treat human analogy reasoning (e.g. 'blind people can do it as effectively as sighted people, so use a tactile method') with great caution, then we are left with a task-specification problem. If we simply look deeper into the task we find *functional* reasons and specifications for what is

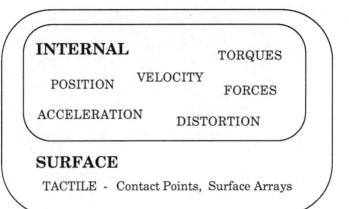

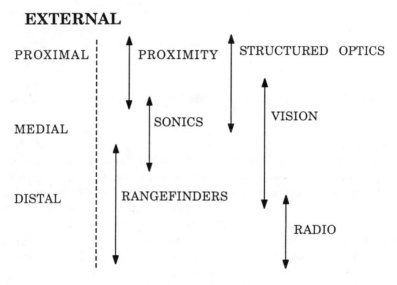

Figure 2.5   Types of sensor

really required. In our example, the task might demand that any pegs involved in an error are to be inspected at a later stage, and are therefore to be placed in a reject bin. In this case the simplest method might be to monitor the motor forces. In another application, it might be desired that a corrective movement is made and a second attempt carried out. The nature of this correction will supply valuable information that further defines the functional role of the sensor. Perhaps, in this case, the force sensing wrist would be the most effective. The point is that by concentrating on the task *first* we discover the functional requirements and can then descend to implementation methods and practical details. Figure 2.6 shows some of the criteria that can be considered at this task function level. In selecting sensors at this level we can weigh their functional relevance and effectiveness before fixing on particular methods or technologies.

Sensing the world 19

> Location and orientation of parts
>
> Recognition of part types
>
> Detection of reject parts
>
> Analysis of spatial relations between parts
>
> Check correct manipulation of parts
>
> Check gripper/tool operation
>
> Work space intrusion detection
>
> Insertion monitoring
>
> Assembly verification
>
> Assembly test operations

Figure 2.6   Task requirements features

Taxonomy Level	Examples
Task Requirements	slippage detection size confirmation inspection defect testing
Modality	vision force tactile
Sensor Attributes	output complexity discrete/continuous variable imaging/non-imaging local/global
Operational level	size accuracy cost
Mechanism level	switching devices inductive sensors vision imaging

Figure 2.7   Sensor taxonomies

We can now list the taxonomy levels described above in top-down order as in Figure 2.7. We should attempt to classify our sensing needs at the task level first and gradually clarify our requirements by moving down through the levels, taking each into account. In this way the practical aspects are always decided within the framework of more abstract task decisions. The advantages of this top-down approach are that sensory needs are properly defined and sensor selection conflicts are resolved in terms of task performance rather than in terms of particular sensor properties. An additional spin-off is that holes in the sensory spectrum are identified and so might stimulate directed research and development. Examples of this could be texture sensors for surface wear or contamination sensors (e.g. grease).

This section has offered more problems than solutions, because such taxonomies are not yet properly developed. There remains much research to be done on the role of sensors in industrial robotics. Eventually enough research effort will be invested in this area so that some research groups will tackle these issues and offer rigorous guidelines and selection procedures that will be very helpful to the system designer. They will prevent similar situations being 're-discovered' and provide important generic carry-over between applications. Until these methods become available, we should try to be impartial, systematic and structured in our selection process. This would be a great improvement over intuitive design and *ad hoc* selection in this multi-dimensional problem area.

2.2 Sensor deployment and integration

Now suppose we have chosen our sensors for our particular robot application and the agonies of selection are behind us. The next job is to incorporate the various devices into our system and arrange them so that they operate and interact in a mutually supportive manner. Sensors rarely operate in isolation because their sensory fields often have broad or diffuse boundaries. It is not easy to ensure that a pressure sensor will only experience the desired stimulus without interference or that a vision sensor will only 'see' a restricted range of objects. Consequently, sensory fields will frequently overlap in some way, especially if a high degree of coverage is aimed for. Consider different manifestations of sensory overlap:

Multi-modal overlap — it is common for two quite different sensory systems to be able to detect and report on a particular event. This has already been seen in Figure 2.1 and is quite easy to produce in practice.

Intra-modal overlap — this occurs when two sensors of similar type have fields that share a common region. The overlap can be simple and direct, as in two cameras that view the same work area, or it may be created through more indirect means, as in internal force sensing in a robot arm where the arm configuration provides a coupling between different sensors.

Redundancy — this is the case of deliberate overlap where one sensor duplicates the field of another sensor. This is usually designed to give high reliability in the sensing system and allow a degree of failure tolerance. Extra controls must be added in order to deal with situations where sensors disagree or provide conflicting data. As there is a considerable amount of experience of such techniques, and a large literature on redundancy design and control, it is relatively easy to adopt some of the standard methods in critical applications. Nearly all such schemes use weighted evidence methods or majority switching procedures.

It seems that overlapping sensory fields are very difficult to avoid in intelligent systems. This is because a high degree of sensory cover is required. The alternative would be very restricted sensing with large gaps in the coverage of task events. Such an isolated sensory scheme would operate like a series of independent individual component sub-systems rather than as an integrated and flexible whole. However, we shouldn't view this necessary overlap as being an area of difficulty, as it may well provide just those features that are desirable in a flexible sensing system. It is well known that sensory overlap in biological systems is extremely valuable and efficacious. Humans can perform tasks with or without a whole modality (e.g. blindness) and there seems to be considerable potential for high level transfer of data between modalities when one system is damaged. At a lower level, intra-modal overlap is extensive in biology. Most visual, tactile and surface sensors have high overlap ratios. In the laboratory, it is usually not possible for a subject to identify the precise location of an isolated stimulus on a single receptor or nerve, and yet when several receptors are activated by a single stimulus due to the normal field overlap, the subject's accuracy improves considerably. This is a very strange state of affairs in engineering terms. Instead of one receptor sending one signal, down one wire, corresponding to activity in a single discrete area, we seem to have a population of receptors sending signals due to a single stimulus. It is as though the address of the stimulated area is being transmitted rather than the intensity image. Despite the difficulties of making immediate use of such ideas in engineering systems, it can be very stimulating to draw on such concepts in our search for design principles for intelligent machinery. Clearly such overlap allows a high degree of fault tolerance, both at the lower levels (sensors) and higher levels (modal sub-systems). This is essential in intelligent systems as the sensory processing mechanisms must be reliable, able to tolerate noisy and incomplete signals, and able to recover from damage or compensate for its effects.

The other key feature is flexibility. As the operational needs change during a task cycle so should the sensors be able to supply different data upon demand. It is not going to be possible to anticipate every sensory need and so some degree of flexibility in the sensing system is very desirable. For example, as sensors are moved into different physical areas, so they must compensate for new ambient levels, interference and noise effects. In addition, there will often be unintentional sensory overlap, of the kinds described above, often referred to as 'cross-sensor coupling', because there will be certain signal variables that seem to be 'connected' in some way. The sensors might also be required to reconfigure their processing arrangements to meet some new request. For example, if the main control system thought it had detected a new batch of components, it might order a sensory system to search for a newly-defined feature or pattern. Such ideas will be useful in opening up our minds to the requirements and design of really flexible systems.

Given that overlap will be inevitable, it seems that we now have to face an integration problem of horrendous complexity. Sensors no longer operate on their own, but must cooperate with others. Results will be produced collectively by patterns of agreement or delegated feature processing. The relationships between sensors are now dynamic and driven from task demands. Clearly we must introduce some form of organization in order to control this potentially unmanageable complexity. Once again we will see the importance of the task in prescribing the role of sensors and the style of their control and coordination.

2.3 Control and coordination

With a large array of sensors incorporating considerable overlap and cross-modal coupling, it is going to be impossible to process the large volumes of data produced without some form of selective structure. If we simply activate all the sensors and collect all the resultant output we merely pass the buck on to the higher processes which must sift through the mass of mostly mundane signals looking for the salient events. A much more selective organization is needed where sensors are active only when their task relevance is high. As the robot moves through the task the sensors should be activated in a highly dynamic manner. Critical operations will be sensed by highly relevant sensors and yet other stages will have less attention. As the task progresses, so the locus of the sensory regime moves in harmony. The criteria for matching sensors to task includes *relevancy*, as mentioned before, and also the *expectations* of the relevant sensors. Expectations are statements about anticipated signals and can define the acceptable and unacceptable performance of the task. We define a sensory *signature* as being a description of the boundaries of expected signals during an event. Now consider a sample of coding for a task. Figure 2.8 shows a small section of a robot program (in a simplified pseudo-code). Assuming that a range of sensors are available, a possible sequence of active sensors is shown for the given task. It is obvious that jaw position, for example, is significant during the OPEN JAWS operation, and is not so important for many other actions. Likewise, vision and proximity, which have been used prior to OPEN JAWS, are no longer needed during the OPEN JAWS operation. In fact, they would introduce confusing and unnecessary input. Clearly, sensors should be turned off as well as on. This implies a great deal of control, as sensors are now constantly being enabled, activated, monitored and de-activated, not to mention large amounts of general processing of sensory data.

However, the benefits are that enormous amounts of computer time are saved in not processing irrelevant data and that the selectivity in the sensory structure provides a clear

Task Program	Active Sensors
Locate (item at) X	vision
Move above X	proximity
Open (jaws)	jaw position
Move (to) X	wrist forces
Close (jaws)	jaw force, jaw position & tactile fingers
Move above X	wrist force & tactile
Move (away to) Y	proximity & tactile

Figure 2.8   Sensory regimes

focus of attention for the current activities. A parallel is seen in human cognition when our attention is directed towards some particular activity or stimulus while all other inputs are ignored. This facility allows us to read while travelling on buses and trains and to listen to a single conversation in a noisy room. Changes of attention control are seen when, for example, a car driver pauses in his conversation during the negotiation of difficult sections of road. In terms of robotics, this attention focusing can be specified as the current sensory context for an operation, that is, the range and scope of the currently active sensors. We will define the attention set as the set of sensors being monitored at a given point on a task cycle. The attention set changes its members according to their *relevancy* at different stages. This context structure proves to be very useful in diagnosing failure events as it provides further knowledge about the task in hand. An attention set could be wide, e.g. monitoring several related events, or narrow, e.g. concentrating on a critical situation. This width of attention corresponds to the degree of concentration. In later chapters we shall see examples of knowledge-based structures that are being developed to handle such dynamic sensory regimes.

Another aspect of sensor control is the mode of use. Sensors are used for monitoring but there are different ways of defining the monitory role. There are at least four distinct modes:

- *confirmation mode*, for checking that an expected event has indeed happened or has produced the right results
- *alarm mode*, for signalling when an unexpected event has occurred
- *probe mode*, for the interrogation of an event and the collection of data
- *tracking mode*, for active monitoring of a continuous event or process.

Confirmation mode involves checking out sensory results against our expectations of those results. This is a common task in robotics and most of the sensors in the active attention set will be operating in this mode. Signatures can be set up to define the expectations.

Alarm mode, on the other hand, concerns sensors and events outside of the active attention focus. These are signals (usually simple, e.g. binary) that indicate a new condition or state has been entered and trigger a switch of attention to a new focus. Consequently, alarms are not continually re-configured as in the dynamic attention set, but are enabled or disabled by explicit commands. These are often called *demons* and are implemented either as hardware or software interrupts. A demon is given a signature for some condition (e.g. power overload) and can then be forgotten. The demon 'lies in wait' from then on (until it is reset or removed) and 'pounces' if and when it sees the specified condition occur. This mechanism takes the onus of constant checking away from the task programming system and results in a cleaner, higher-level model of the control task.

Again we can find an analogy from psychology. The 'cocktail party effect' is the well-known situation in a noisy party when it is hard to hear anyone properly and yet when your own name or any other personal detail is mentioned, perhaps far away in the room, your attention is immediately captured. It is interesting to experiment with this: shouting 'fire' is not nearly as effective (because most people do not have real expectations of fire?) as shouting 'police', in which case any drinking drivers suddenly seem to turn less jovial!

Probe mode covers a wide range of sensory interrogation activities. Simple probing could be used to locate the height of an object while more elaborate monitoring could return measurements, inspection data, classification results and, most general of all, descriptions produced by pattern recognition methods. Some sensors in our robotics applications will operate in probe mode at selected points during the task and will be controlled by the attention and signature settings. The more general processing issues are discussed in the next chapter.

Finally, *tracking mode* is a monitoring role that is becoming increasingly important. Tracking of seams and cracks is the obvious example as seen in robot welding and inspection systems. Such adaptive tracking can be very useful in automatically compensating for minor uncertainties in the location of the seam. It is far more efficient to set up a sensory system to handle a general family of minor variations than to program in each member of the family in a non-sensing system. This type of activity has its roots in control theory, as such close coupled sensing and movement can often be efficiently implemented using feedback methods drawn from the relevant mathematical theory. However, as robotic assembly becomes more of a reality, so does the need for tracking actions under controlled forces or pressures. Robots often have to follow surfaces while maintaining some specified values of variables such as distance away, surface force vector or wrist forces. These are called 'guarded moves' and can be implemented in most modern robot programming languages. Such facilities are very important for assembly where manipulations under controlled forces and spatial configurations, such as alignments and contacts, are necessary. Tracking operations are often controlled by dedicated, autonomous sub-systems that report when the movement is completed or when signature errors occur. This allows tracking to be treated as a single operation by the higher processors and thus does not tie up the main computer power inside the tracking loop.

We see now that the way sensors are integrated into intelligent control programs is a complex issue. Sensors are not simply extra 'input peripherals' but have a pervasive effect on the design of the whole system. A navigation example serves to illustrate this. Consider the control of a movement from A to B. One extreme method would be to gather all the sensory data first, then compute the details for a path, and then execute the path without further input. This is the dead reckoning approach and is the natural style for symbolic computation problems. The other extreme would be to use an error-correcting feedback-control loop which continually adjusts its path based on incoming sensory data. This is usually a real-time situation and, as there is little scope for extensive symbolic computation, dedicated electronics are often employed for its implementation. In robotics, we experience both of these cases but the general sensory control problem lies somewhere in the middle. We want to plan ahead on the basis of sensory data as far as possible but we must accept important sensory input whenever it occurs. The problem is that we usually *do not know* when the relevant sensory data will arrive. This is the main reason why software attention and integration controls are so valuable.

2.4 The importance of constraints

We have seen that selectivity helps in the management of the data collection problem. Sensors in the attention set are carefully monitored while others are only perceived when alarm conditions occur. Attention can be viewed as a sort of internally-imposed constraint on sensory input. The value of such constraints is considerable in reducing sensory complexity. We can now ask, are there other methods of constraint that will help still further? Attention is an internal constraint, so let's look at the external world in front of the sensors.

Suppose that a number of washers are to be measured across their diameters. The washers are presented lying flat and are carried on a moving belt past the sensing station. Consider two possible options — a two-dimensional array scanner (the mechanism might be visual or tactile, but this is not significant) and a linear scanner. In the first case an image of the object is captured and then internally processed to determine the diameter. In the second case the linear scanner is used to record the length of line it detects as the washer passes by. The maximum length is the diameter. Clearly, in the second case the cost, complexity and processing time will be considerably reduced. Only a simple logical hardware register is required whereas for array imaging a small computer will be needed with attendant equipment and complexity. This saving has been achieved both by constraints on the physical sensor hardware, i.e. linear rather than array technology and, more significantly, by the use of a constraint on object shape: the assumption of circularity. In terms of the *functional requirements* of the task, both systems are equivalent; they both report the diameters of washers with reasonable performance parameters. However, as soon as we wish to alter anything we soon feel the restrictions imposed by our constraints. If, for example, the washers were stationary during the inspection process, then the linear method would fail completely. Also, if non-circular objects were used then the diameter recorded would depend upon the relative orientation between linear sensor and object.

It is useful to assess the relative severity of imposed constraints. Some constraints are much less limiting than others. The linear nature of our sensor is not really a handicap *provided* that object movement is present. In fact this combination can offer the full functional equivalence of an array sensor. Our real constraint is in the limitations of the processing hardware. If we replace this with a storage array and clock out a series of line scans from the linear sensor then we achieve an image store that can be used to process all sorts of measurements of the object. Of course, we don't need to provide such power in many cases, and here lies the simplicity of the first solution with its attendant savings of cost, reliability and ease of use. This trade-off between potential usage (i.e. flexibility) and simplicity is an essential feature of the use of constraints. AI has frequently opted for maximum power and flexibility by trying to avoid all constraints. However, the problems are so difficult that research has often been forced into more tractable areas, at least until more is understood about the problem domain. An example of this is scene analysis work. Programs that try to figure out where trees are in natural outdoor scenes have a much harder time than versions that look at special laboratory views of regularly-shaped objects with smooth, even surfaces, with no cracks and no shadows. Fortunately, industrial scenes are far more constrained than views of natural objects, being mainly full of man-made objects and artifacts. It is just these features and assumptions that we can capitalize on when designing intelligent systems for industry. Thus we can view constraints as helpful features rather than frustrating obstacles to overcome. Later chapters will reinforce these ideas.

2.5 Summary

1 Sensor selection is a big problem. Scientific and systematic design methods are needed to aid the development of high-quality robot systems. Intuitive methods only work for

short-term or fixed systems and will be inadequate for flexible manufacturing systems. More research is required in this subtle and easily-overlooked area.
2 Task-oriented design methods are more powerful and comprehensive than sensor-specific design methods.
3 Sensors can be used in several quite different modes.
4 Sensory field interaction is important for intelligent systems and should not be viewed as undesirable.
5 Human analogies are useful stimulants for ideas but should not be followed dogmatically. The danger is that by building anthropomorphic solutions we may overlook far simpler and more elegant solutions.
6 Robot sensors are not simply 'input peripherals' but need the support of highly complex and dynamic software control environments.
7 Attention-focusing mechanisms provide a useful means of handling the sensory information explosion and establishing relationships between sensors and their role in tasks.
8 Constraints can play an important role in simplifying any given sensing problem. All types of constraint should be considered and used whenever the simplicity/flexibility trade-off allows.

2.6 Further reading material

There are many manufacturers' catalogues and data sheets giving sensor characteristics and operational and performance information. There are also many textbooks on sensors, transduction and measurement technology. Examples are *Robot Sensors and Transducers* by S. R. Ruocco (Open University Press, 1987) and the two-volume collection of papers entitled *Robot Sensors*, edited by A. Pugh (Springer, 1986). The journal *Sensor Review* gives information about new sensor developments.

There has been little work on the fusion of sensor data. A paper by T. Henderson and E. Shilcrat, 'Logical sensor systems', in the *Journal of Robotic Systems* (1984), volume **1**, number 2, pp. 169–193, deals with a systematic approach to sensor integration. The book by H. Durrant-Whyte, *Integration, Coordination and Control of Multi-Sensor Robot-Systems* (Kluwer-Nijhoff, 1988), looks like a promising text. Another source is *Spatial Reasoning and Multi-Sensor Fusion: Proceedings of the 1987 Workshop*, edited by A. Kak and S. Chen (Morgan Kaufman, 1987).

For robot motion control and monitoring, the paper by T. Lozano-Perez, 'Robot programming' in the *Proceedings of the IEEE* (1983), volume **71**, number 7, pp. 821–841, gives a good discussion of the requirements for sensors in robot control.

Chapter 3

Artificial sight

He that has one eye is a prince among those that have none
Thomas Fuller

Sensors which are organized as imaging systems are so significantly different from other types of sensor that they merit special attention. This chapter describes some of the main features and techniques found in industrial imaging sensors. Vision will be used as an example throughout but the ideas apply equally to any form of imaging sensor, be it tactile, thermal, ultrasonic or even arrays of switches. The distinguishing feature is that many identical sensing elements are arranged in a uniform (usually rectangular) array.

3.1 A salient characteristic

Imaging sensors employ either a large number of identical sensing elements or some form of single scanning sensor. In either case, the resultant output is an array of sensory intensity values, called an *image*. The actual values can be binary or many valued (grey scale). One major feature of such sensors is the enormous amount of information they can generate. For example, to give reasonable quality pictures, the output of a video camera can be digitized into 256 x 256 picture elements, called pixels, with an 8-bit grey scale (giving 256 levels of greyness for each pixel). However, to store one such picture requires 65 536 bytes of computer memory. Even to store a crude binary image (grey scale of two levels) requires 8192 bytes of memory. For this reason it is very important to process the images *as soon as possible* in order to extract the important and relevant data for the application and avoid storing massive quantities of raw data. In case these figures seem too high, consider the information content in an ordinary television picture. You would need around 1 million bytes to store the information in one full frame of monochrome, and there are 25 or 30 of these every second! Colour television, of course, contains even more information! We often don't appreciate the amount of data involved when we use video tapes and televisions; it is only when we have to digitize images that we see the complexity inherent in this media.

Clearly, much of the information in a video picture could well be redundant data or of low significance at any given time. However, we don't want to lose information about the critical area of an image where a vital sensory event has occurred. So, once again, we are faced with a selection problem. How can we filter out all the irrelevant image data and focus on the critical variables?

3.2 Computer vision applications

The answer to the above question must be the same as for previous sensor selection questions, that is: 'it all depends on the needs of the application and the role of the sensor in the task'. There are many application areas for vision and image-processing systems, including medicine, biology, cartography, aerospace and undersea work. There are also many fields where sophisticated image-enhancement methods are used to restore images, for example, in communications, telemetry, satellite remote sensing and weather prediction. Our particular concern, however, is industrial automation and we are interested in imaging applications in robot assembly, robot materials handling, and industrial inspection of all kinds. We can also restrict our application area to fixed work cells and stationary manufacturing machines. Notice that mobile robots and vehicles incorporating vision sensors have quite different requirements. Mobile systems not only have to contend with

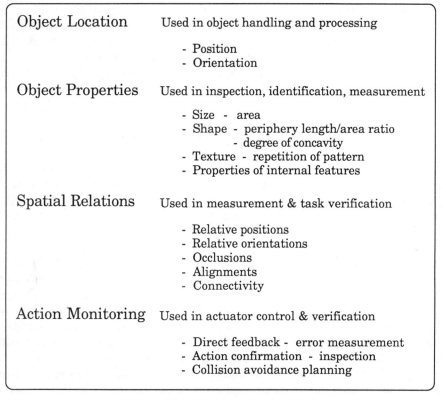

Figure 3.1   Industrial applications for vision

Artificial sight

object recognition but have 'higher-level' problems involving spatial mapping, interpretation and navigation. These often involve scene analysis, depth perception, and stereo and motion interpretation. Fixed production robots and most intelligent factory machines, on the other hand, are engaged in identifying, inspecting and measuring particular objects in their work space. This object-centred approach is generally easier to manage than the analysis of complex dynamic events in large-scale environments. We take the view that research on mobile robot problems is of value to industry, but is long-term work which will take some time before it is available for commercial exploitation.

3.3 Industrial requirements

We have already discussed the task requirements for sensing systems and Figure 2.6 illustrated the main classes of industrial application for sensors. For vision, these break down into four different categories as shown in Figure 3.1.

Object location involves the exact positioning of parts by finding out where their centres and their major axes lie in space. This can provide sufficient data to identify grasp sites for a robot gripper. *Object properties* are all those intrinsic features such as size, shape, texture, surface finish, and other, more contrived, measures. These are often used as parameters in recognition and inspection processes. *Spatial relations* concern the distances between parts and whether parts are correctly organized in assemblies. Finally, manipulator, and other machine, actions can be visually *monitored* and this can include all forms of assembly action, inspection of system states and examination of potential collision situations. With these general requirements in mind we will explore further the nature of vision systems and discover what they can offer in the industrial environment.

We must remember that the performance variables are very important; for instance, costs must be low and the speed of processing must be relatively high. Any system that takes more than one second to recognize an object will not be well received in a factory environment. A vision system must also be flexible in that it is easy to reconfigure for different objects. This will usually be arranged through a customized user interface for the factory floor rather than by being re-programmed by vision specialists. Also, it must be robust; this is one of the most important requirements. Unfortunately, many AI systems are rather 'brittle' and liable to break down when tested near their performance limits. This will not be acceptable in an industrial environment, where 99% reliability is desired over the full operating range of the device.

3.4 Research fields

Let us now look at the main research fields in computer vision and introduce some basic concepts. Then we will review some common techniques, look at industrial and commercial systems, and then, in the next chapter, we examine the special problems of interpreting visual and other images.

Figure 3.2 shows the three main research fields. First of all, *image processing* is the area of study that concerns the transmission, storage, enhancement and restoration of images. A typical system will take an input image, apply a selected process, usually to improve its

IMAGE PROCESSING

- results in new

PATTERN RECOGNITION

- results in image

IMAGE UNDERSTANDING

- results in image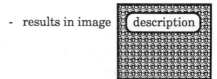

Figure 3.2   Three research fields

quality, and then output the new image. Applications are in such fields as remote sensing from satellites, weather forecasting, and all forms of image transmission and communication systems. There are some AI techniques involved in this field, but we will not pursue these, as our main concern is not image processing but image analysis.

The next stage, *pattern recognition*, is an area of study that involves the interpretation of an image in terms of a pre-defined classification of patterns. Good examples are the recognition of hand-written characters, cheque-book signatures, car number plates, fingerprints, or microscope specimens. This field has two very significant characteristics: (1) the objects generating the image are usually two-dimensional and (2) the objects either have formal definitions or are declared as variations on a prototype. The problem is to determine which one of a set of symbol or character patterns has been seen and analyse the likely probabilities for different classes. There is a considerable literature on pattern recognition with a large body of algorithms and statistical techniques.

Finally, the more difficult area of *image understanding* deals with the *perception* and *interpretation* of three-dimensional scenes. This is more difficult because nearly all imaging systems have only two-dimensional receptors and thus depth information is often missing. The problem then is one of interpreting three-dimensional objects from two-dimensional data — a potentially serious situation in which vital information might be lacking. These systems are concerned with symbolic representations and high-level understanding of configurations of surfaces and objects in space. They often attempt real-life scenes, such as houses and trees. The output is a description of the scene or image rather than a simple classification result as in most pattern recognition schemes.

Artificial sight

Showing noise from imaging
and digitisation processes.

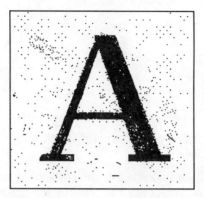

Figure 3.3   Example input character

3.5 Pattern recognition

We first consider the ideas involved in pattern recognition because the easier problem of two-dimensional interpretation serves as a good introduction before moving on to three dimensions. It is also important to be familiar with basic pattern recognition concepts as many techniques widely employed in various other areas, such as scene analysis and speech recognition, have historical roots in the earlier two-dimensional methods.

Typically, the input material for two-dimensional pattern recognition will be flat, two-dimensional objects, for example, written characters, graphical symbols, microscope images, X-ray images, or aerial photographs. These are digitized by photo-cell arrays, TV cameras or scanners of various sorts. The image will have some digitization error, and its complexity of rendition may vary, ranging from photographs to line diagrams. Figure 3.3 gives an example of a single character image for recognition. Noise and image irregularities are very common in most photo-imaging methods.

There are four basic categories of technique in pattern recognition. The first one, called *template matching*, simply consists of a matching between a standardized model of a character and the input. The template is often a near literal copy of some preferred character. This is quite suitable for recognizing standard printed fonts that are used in printing or typewriting. Frequently some normalization will be performed beforehand, in which the character is shifted and scaled in size or rotation. Simple thresholding techniques can be used to determine which of two similar characters is the more likely. All such direct comparison methods can only deal with minor variations of input images. For instance, each of the characters shown in Figure 3.4 would require its own template as there are wide variations between the characters.

Perhaps the most important pattern recognition method is that called *feature analysis*. Figure 3.5 illustrates the basic method. A series of different image features are computed by pre-programmed procedures which scan over the raw image. These features can be highly

Figure 3.4 Images of a single letter

specific, such as different types of line end or corner, or they can be global features, such as the ratio of bright areas to dark areas. The features are selected by the designer of the system and should be appropriate to the recognition task. The outputs of the feature detectors are quantities which represent the amount of each particular feature present. A

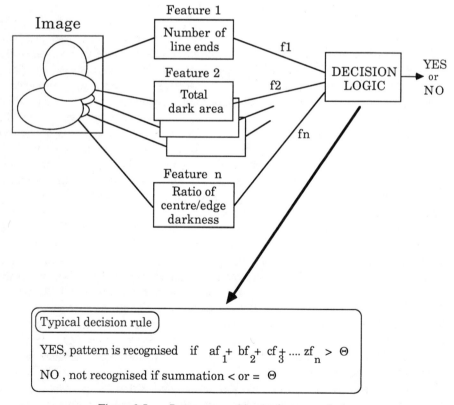

Typical decision rule

YES, pattern is recognised if $af_1 + bf_2 + cf_3 + \ldots + zf_n > \Theta$

NO, not recognised if summation $<$ or $= \Theta$

Figure 3.5 Pattern recognition by feature analysis

Artificial sight

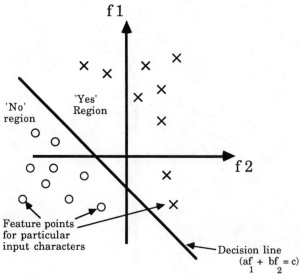

Figure 3.6 Feature space

decision classifier then weighs the outputs of these feature detectors — a decision is based on whether the required group of features exists for the particular character to be recognized. It is common to use numeric weights for each feature and to use a linear summation in order to decide whether the weighted feature response is above a particular threshold value. Thus, the system shown in Figure 3.5 is able to give a binary answer for the presence of a particular pattern or character to be recognized.

A very simple system with just two feature detectors will illustrate all the essential concepts (see Figure 3.6). The two features provide the axes of a two-dimensional plane known as *feature space* and any observed image can be represented by a plot point in feature space. If the examples of the images cluster into groups then it is relatively easy to design a decision line which will separate the two regions, thus providing the recognition process. In Figure 3.6, the decision line is based on a linear weighted sum of the two features and there is a clear separation between the two groups of feature points. More difficult plots of features might require more complicated decisions which will involve curved lines or even variously shaped enclosed regions. In more realistic examples, there will be many different features and so feature space becomes a multi-dimensional space and decision lines will become decision surfaces representing partitions of the space. There will also be more decision lines or surfaces as more characters are required to be recognized. Each particular pattern will require at least one decision line or surface.

There are other methods of determining the class of input patterns and Figure 3.7 illustrates the *nearest neighbour method*. In this situation each new pattern is compared with its nearest neighbours in feature space. In this way the new pattern is recognized as an example of the cluster to which it is nearest. The nearest neighbour decisions may be based on the single nearest point (i.e the point giving the smallest value of $d1$, $d2$, $d3$...), or on a weighted average of the clusters by some more complex recognition algorithm.

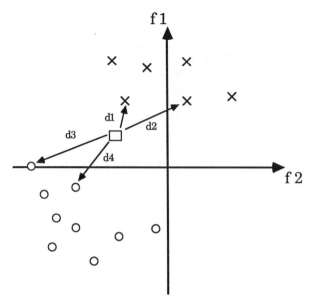

Figure 3.7   Nearest neighbour strategy

One reason why feature space methods have been popular in the past is that *learning algorithms* can be incorporated into such weighted methods. Figure 3.8 shows the same feature space as for Figure 3.6, but a series of new inputs have been received and the recognition system has been instructed to learn these as new examples of the pattern and counter-pattern. Unfortunately, some of these points lie on the wrong side of the decision line, and so it is necessary to compute a new decision line which maintains the separation of the original cluster but makes sure the newly experienced images are on the right side of the line. Because the weights attached to the features simply represent the slope coefficients of the lines, it is possible to design an incremental weight adjustment algorithm that corrects the decision line slope to take account of new examples. This method forms the basis of many learning processes in pattern recognition.

However, there are serious difficulties with these methods because patterns may not always be linearly separable. In this case, the learning method will have to handle much more complex decision lines or surfaces and we do not have satisfactory and efficient techniques for this. There have been many variations on such feature space methods and there is a long history of this type of recognition process being used in all forms of learning and perception research.

The third category is called *relational analysis*. In this approach, feature detectors are arranged to measure parameters of the geometry and relationships between the components in the image. Thus, a given character might be seen as consisting of several lines, at certain distances apart, slanted at certain relative angles and of certain relative lengths. These measurements would then be compared in the same way as for feature analysis by using a suitable decision rule. The difference is that the relations *between* image features are the important variables here, rather than the *amounts* of each feature present.

Finally, there is the *syntactic method* of pattern recognition in which a picture grammar is

Artificial sight 35

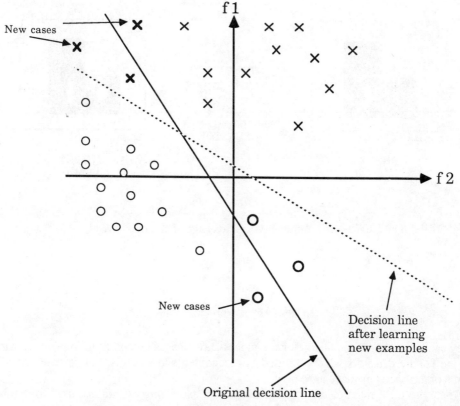

Figure 3.8 Learning by weight adjustment algorithms

used to parse the image into a suitable format for decision making. This is analogous to language parsers. Pre-defined primitives are detected, and then structured into classes to determine whether the image is of a particular acceptable form or not. This method produces a description of the input rather than a classification response. The advantage of descriptive methods is that new images that have not been seen before can still be processed and described in some way. Every image is described, even if it is a hybrid of other patterns. By contrast, classifiers will either reject any strange symbols (by placing them in a special 'reject' class), or, worse, will mis-classify them, thus giving recognition errors.

3.6 Basic techniques — Local operators and segmentation*

Nearly all image processing systems employ some form of localized processing techniques. These local methods have become a major part of computer vision research. This is because

*Section 3.6 contains technical detail of some standard image-processing techniques and some readers may want to skip over this. This section is included to show that useful (although limited) methods have been developed for certain industrial situations. This provides balance because much of the later material deals with the problems and difficulties in AI applications and tends to be cautious about the availability of powerful image understanding software.

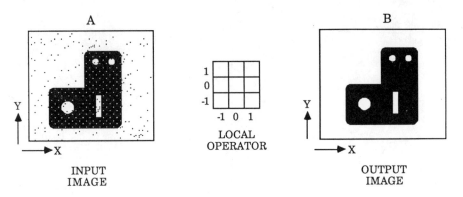

The local operator output is the average of the mask pixels:

$$B_{x,y} = 1/9 \sum_{i=-1}^{1} \sum_{j=-1}^{1} A_{x+i,y+j}$$

Figure 3.9   Noise removal

they can be used both to enhance or pre-process raw images as well as to implement many of the feature detectors which are used in recognition and analysis. Let us now consider some of the operators that can be used.

There are three main classes of local operation: *smoothing operations*, used to remove noise and distortion, *gradient detectors*, which can detect either gradual or sharp transitions in image intensity, and *thresholding* in which the image is converted into a binary version.

Figure 3.9 shows noise being removed from an image. A small mask is used to examine the values in a localized region of the array. A popular mask design is to use 9 cells in a 3 x 3 square, although other formats are frequently used. A logical or arithmetical rule is used to compute an output value which is assigned to the output image. The mask is then moved to other positions, usually in a sequential scan, until the whole image has been processed. In this way, any desired processing function can be scanned across an image to produce a new output image. This method of processing one image into another by local methods is fundamental to computer vision.

The noise operator shown simply detects isolated high spots in the image where a pixel is surrounded by significantly different values. It changes the centre point to the average of the surrounding ones and so removes local noise, sometimes called 'salt and pepper noise', because it looks like a sprinkling of dots on top of the image. Notice how in Figure 3.9 the regular 'textural' pattern of dots on the object has also been removed by this operator. Averaging over larger masks produces more extreme effects of this type which actually defocuses the image and smooths it in a blurring sense. This simulates the defocusing action of lenses. Figure 3.10 shows the effect of a smoothing operator on a sharp edge.

The operator illustrated in Figure 3.11 is a gradient detector. The input pixels are multiplied by the operator weights and, if the summation of these is greater than a given threshold, an output signal is generated for that operator position. The operator mask is

Artificial sight

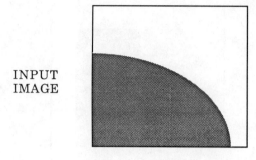

INPUT IMAGE

Large averaging operator Centre square indicates output position. Output is average of all input pixel values.

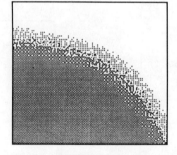

OUTPUT IMAGE

Figure 3.10   Smoothing or averaging operator

designed so that a sudden change from one intensity to another will result in a high summation value which is recorded in the resultant output image. This output image shows a series of points where the input image had high gradients of intensity; these will often form lines giving the boundary of image features. There are many designs of gradient detector varying from attempts at generalized versions, such as the one shown in Figure 3.11, to more specific edge operators that look for a particular shape of edge. Quite frequently there will be a requirement for several different orientations of operator as a single edge operator may only be able to detect one type of edge orientation. Edges are usually the interesting features in an image, and this operation picks out the highlights or silhouettes that tend to be the important parts.

Notice that smoothing and edge detection are inverse processes. Smoothing is an integrating effect that tends to even out and blur the image, while edge detection exaggerates any sharp image features by differentiation. This is why edge detectors are sensitive to noise and usually produce many spurious edges in an image.

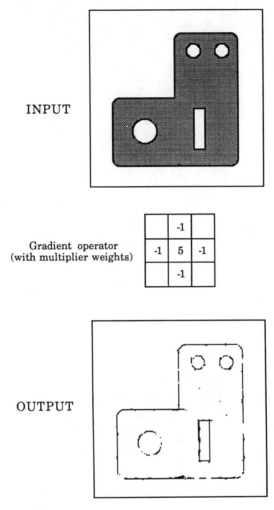

Figure 3.11 Gradient or edge detection

Edge finders are really only examples from a more general class of feature detectors. We can design a special mask for any desired feature that we wish to detect in the image. By 'pushing' our special operator around the image we will discover if the feature is present, and where it lies. However, this approach must not be taken too far or else it becomes self-defeating by over-specialization. If we insist on constructing idiosyncratic detectors we will end up writing 'grandmother' finders or 'yellow Volkswagen' detectors. This argument leads to an infinity of detectors with no generality. Instead, we try to design more general purpose operations and build a library of primitive operators that can then be called on to allow us to construct powerful and flexible analysis functions.

The third class of local operator, thresholding, is really an extreme form of gradient detection. Each pixel is set to one or other of a binary value according to whether its grey level is above or below a pre-defined threshold setting. This has the effect of producing

Artificial sight

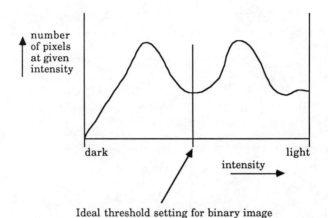

Ideal threshold setting for binary image

Figure 3.12 Global threshold determination

silhouette type images by separating the black areas of the image from the white. If the image is of high contrast then the segmented regions, called 'blobs', often correspond to objects to be recognized or located. The threshold setting can be determined by a prior histogram analysis of the image, during which the number of pixels at each intensity level are counted. Some images produce a bimodal histogram distribution (see Figure 3.12), and so the minima in the valley becomes the natural threshold to use for the separation of light and dark. Such global thresholds can be set up either by human operators or by an automatic histogram analysis of the image before processing. Another method is 'adaptive thresholding', where the threshold setting varies locally depending on the background of the region being processed. This can be useful in the many cases where the distribution is not bimodal.

There are many other designs of mask for local operators. They can carry out a wide range of different techniques. For example, size can be measured: an operator can be designed to measure the length of a visual segment or the area of a region. Other masks can be designed to shrink the image or expand its size. Noise removal of various forms is possible. Texture (i.e. repetitious patterns) can be detected and decoded, and images can be enhanced in various different ways. Figure 3.13 shows an example of several operations applied sequentially in order to detect the holes in an engineering component. Figure 3.14 illustrates another approach where a purpose-designed feature detector is used to look for certain shaped holes. Following this type of image processing, some decision logic could analyse whether sufficient number of holes were present, whether their locations were satisfactory and decide if any other features were also acceptable. If these were all present then the decision logic may signal a satisfactory component.

Perhaps one of the most frequent requirements in industrial object inspection is finding the location of an object. This can be achieved by detecting its surrounding perimeter. The process is known as *segmentation*, because the image is to be broken up into coherent areas of similar intensity. These segments should define objects by locating their boundaries and will also pick out significant object features. The first stage involves an examination of changes of gradient and so local operator edge detectors are frequently employed. In the early inexpensive vision systems, thresholding became a popular technique as successful

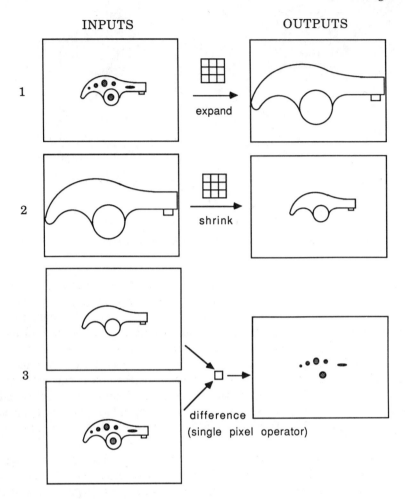

Figure 3.13   Operation sequence for hole detection

thresholding can give results equivalent to gradient detection followed by a segmentation technique. However, gradient or edge detection followed by a separate segmentation process will offer more control and scope for better results.

There are two main methods for segmentation: *edge finding* and *region growing*. The edge finding technique uses local line-finding operators which generate a picture of all the lines that can be found in the image, as in Figure 3.11. These lines are then joined up by continuity algorithms which try to establish connections until a region has been defined by a continuous boundary line. Region growing is a quite different segmentation method in which regions are formed by merging areas of similar intensity. In this technique local neighbourhood processing is used to grow selected seed points outwards until the boundaries of different regions meet. In a particular image, patches of regions are expanded in size by merging pixels that have the same or similar intensity values. These regions grow outwards until they meet boundaries between other areas of significantly different intensity.

Artificial sight 41

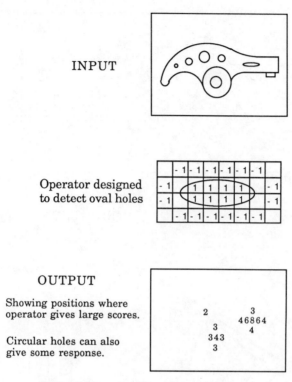

Figure 3.14 Special purpose operator

When adjacent regions are too dissimilar in value, the merging process stops and segmentation is complete.

Region growing can produce problems of ambiguity, depending upon the order in which regions merge, and these must be resolved by algorithmic rules. However, there are problems in line interpretation techniques, as well. Lines may not join up or they may overlap, and both methods have their problems if the image is noisy or badly encoded. These are two alternative techniques, but the end result is intended to be a series of image regions indicating the objects and their relationships. It is interesting to note that lines are duals of regions: lines define the boundaries of adjacent regions and two adjacent regions define a line. The two segmentation processes can also be seen as duals: either edge fragments are grouped into lines, or patches are merged into areas. It is possible that both techniques could be combined to offer their joint advantages.

An example of segmentation is given in Figure 3.15, in which an image has been processed into its various regions and a hierarchical data structure has been created to capture the structure in the image. As each of the different regions is detected, a descriptor is generated and associated with it. A descriptor holds computed properties such as: the area of the region, its centroid coordinates (the centre of gravity), the perimeter length of the region, the number of holes, links to any sibling or descendant regions contained within the region, etc. Such properties are useful in classification and recognition tasks.

After segmentation, there are different ways of representing the resulting region shapes.

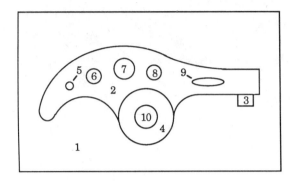

SEGMENTATION STRUCTURE

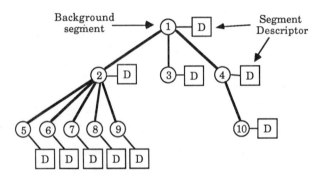

Figure 3.15   Segmented image

These descriptions of shape are either based on area measures or on edge parameters. One edge description method is the *boundary tracking technique*. In this technique, any edge point on the object is first found (by scanning across the image) and then a boundary tracking routine follows along the edge of the object until it eventually retraces its steps back to the beginning. Figure 3.16 shows a diagram indicating the method. The output from such a process will be a chain coded sequence giving the perimeter shape of the object. Another method, useful in fixed raster scanning systems, is *run length encoding* in which the 'run length' of each distinct segment of a scan line is measured and stored. Figure 3.17 shows a fragment of run length encoded image. Each line of the image is represented as a series of blocks of uniform intensity or colour.

The output of chain or run length encoding, as shown in Figures 3.16 and 3.17, indicates how shape information can be compressed into a series of numbers. However, these are only crude representations and have their own limitations. In the chain code case, random access to any point of the image is required in order to follow the edges, and so it is not suitable for a raster scanning system which is the basic method of scanning used in most video systems. Also, only the outline of a shape is represented, and this cannot be rotated or changed in scale very easily without considerable computation. Run length encoding

Artificial sight

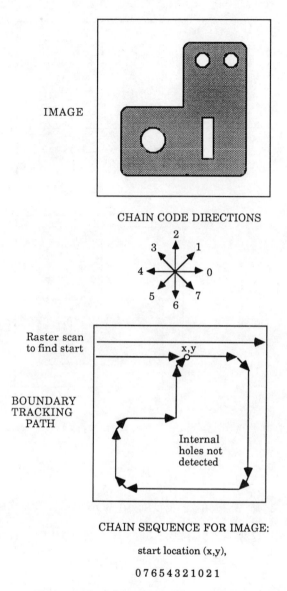

Figure 3.16   Boundary path encoding

contains much the same information but in a different form. This method is very suitable for line scan or conventional television cameras as it allows very efficient comparisons of images, even during the actual image acquisition process! In either case, these two encoding methods have been useful for machine vision purposes as they compress much detail about object shape into a compact representation that can be efficient for simple recognition and checking purposes.

From these kinds of process, we can derive an analysis for a particular industrial component that incorporates many different properties. As mentioned above, these could

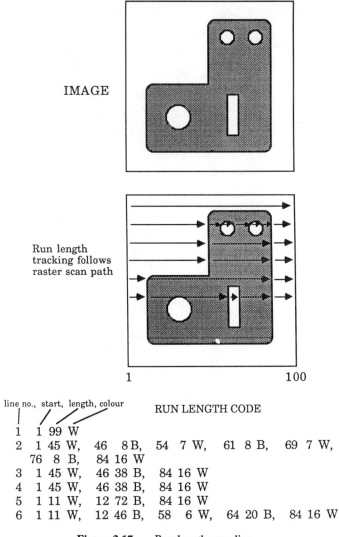

Figure 3.17 Run length encoding

include the area of a region, the perimeter length, the centroid's position, the number of holes, the maximum radii from the centroid, the minimum radii, and many other simple geometric properties. These can be surprisingly successful in distinguishing between different components and faulty or malformed shapes. It is often very useful to have measurements which do not vary with translation or rotation of the object. One such set of invariant features can be computed by a technique known as the *method of moments*. Figure 3.18 shows how moments are calculated. Each moment is the sum of pixel distances from an arbitrary base line. The zero-order moment is simply the sum of the pixels in the image, i.e. the area of a binary image. The first-order moments give the sum of all pixel distances from the base line; if this is divided by the area we obtain the mean pixel distance, i.e. the

Artificial sight 45

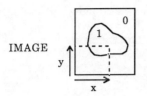

For a digital binary image (background = 0, object = 1) the i,j order moment is given by :

$$m_{ij} = \sum x^i y^j$$

summing over all pixels in the image.

m_{00} = sum of all pixels = area of object

$$m_{10} = \sum x$$

$$m_{01} = \sum y$$

$$\left. \begin{array}{l} \dfrac{m_{10}}{m_{00}} = XC \\ \dfrac{m_{01}}{m_{00}} = YC \end{array} \right\} \quad \text{Coordinates of centroid (i.e. centre of gravity of object)}$$

If moments refer to centroid they can become <u>invariant</u>

e.g. $\overline{m}_{ij} = \sum (x - XC)^i (y - YC)^j$ is invariant under <u>translation</u>.

Using second order moments :
$$m_{11} = \sum xy$$
$$m_{20} = \sum x^2$$
$$m_{02} = \sum y^2$$

Similar formulae can be calculated that are invariant under <u>rotation</u> and <u>scale change</u>.

Figure 3.18 Moment calculations

location of the centroid. Second-order moments can be used to calculate a set of invariant moments that are unaffected by various systematic transformations or distortions. Some moments are even invariant with scale changes due to size increases or shrinking. The importance of the invariant moments is that they can be compared without having to normalize the image to compensate for positional or rotational variation between different views. All these moments can be calculated, with little extra work, at the same time as the area and perimeter features are being computed.

Many of the features described in this section are available in commercial vision processing systems.

3.7 Industrial vision systems

A number of commercial packages can now be bought for vision processing work. These commercially-finished systems may be thought of as first- and second-generation vision

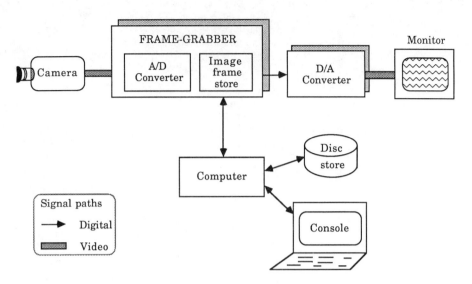

Figure 3.19 Vision system hardware

systems. Figure 3.19 illustrates a typical hardware configuration; a television camera is interfaced to a computer system via a frame-grabbing device. Solid-state cameras have the advantages of robustness, small size and digital outputs, while vidicon cameras are cheaper and give standard video output. The frame-grabber is effectively a large memory buffer that can hold one or more images. The computer is usually a conventional general-purpose microcomputer that can operate on the image to perform local operator processing or other algorithmic processes. Clearly, a conventional computer will only operate in a serial mode, and so the scanning of any local operator mask will take some time. However, by arranging the hardware to be dedicated to this task, reasonable speeds can be achieved. A typical system will have facilities for controlling the camera remotely and perhaps interfaces for remote lighting controls. There may be other links, such as serial communication links, to connect to robots, control computers or manufacturing systems. The operations which will be involved include the acquisition of an image, the processing of image data, and the matching of image features against stored prototypes or models. At each stage, we may have simple solutions or more sophisticated and complex processes. For example, the acquisition process could produce a grey-scale image with 1 byte per pixel (grey scale of 256 levels), or it could simply be a binary image of 1 bit per pixel (grey scale of 2 levels).

The potential power of any given vision system depends upon its software, and there are three types of software system that may be purchased. The first system could be called a development or experimental system. In this case a large library of vision operations is provided and the user can experiment with these to produce custom-built image processing for a given application. Figure 3.20 suggests the range of commands that are available in a typical system. When a suitable sequence of operators has been determined, these can be packaged into macros or other procedural segments of code. The vision procedures can then be used as components in a larger program or downloaded for use on a smaller computer. Although development systems provide good facilities for experimentation and devel-

Artificial sight 47

PICTURE TRANSFORMATIONS	FEATURE EXTRACTION
Aspect Adjust	Binary Edge
Count White Neighbours	Cross Correlation
Expand Grey Areas	Centre of Gravity
Edge Remove	Radius of Curvature
Exchange Picture	Freeman Chain Code
Largest Neighbour	Intensity Lower Limit
Polar to Cartesian Transformation	Median
Picture Expand	Maximum X Value
Remove Isolated Points	Minimum Y Value
Picture Shift	

FILTERS & GRADIENTS	PICTURE ARITHMETIC
Radial Gradient	A Picture + B Picture
Fast and Crude Gradient	Bit Complement
Roberts Gradient	A/B
Sobel Gradient	Minimum (A, B)
High Pass Filter (Laplacian)	Negate
Low Pass Filter	Scale
	A + Constant

IMAGE INPUT AND OUTPUT	HISTOGRAMS & THRESHOLDING
Clear Frame Store	Histogram Plot
Set Up Display Areas	Histogram Equalise
Digitise	Histogram Tabulate
Print Image	Dynamic Thresholding
Write Image To File	Threshold
Set Type of Camera	

Other Facilities are often provided for decision making, program control and debugging.

Figure 3.20   Typical operations provided in a vision development system

opment, they require some expertise to use and there is a long learning time in order to understand the parameters of vision and to appreciate fully its characteristics. These systems have the potential for quite sophisticated vision algorithms but depend upon the user's expertise and experience.

At the other end of the spectrum there are turn-key systems which provide automatic software for recognition and inspection. Frequently these systems incorporate a learning phase during which typical examples of components are presented in front of the camera. During this phase, a wide range of image properties is computed and stored for each example. Then, during an interactive session, a decision rule is built up to determine which of the features are the most useful in segregating the components into different classes. This decision rule is then used during the operational phase when new objects are to be recognized. The features which are computed for these systems include some of those already mentioned — areas, perimeters, the total length of arcs of curvature, the number of arcs, the number of holes, the number of distinct shapes, and so on. The hope here is that if enough features are calculated those important features which discriminate between different classes of objects will be covered. These systems have the advantage that relatively non-skilled workers may set them up, but they have the disadvantage that the software is not readily accessible. Consequently, it is not easy to change the recognition algorithm for major changes of application. Two techniques that have been used in such systems are the nearest neighbour strategy, as described in Figure 3.7, and binary decision trees, using, at each branch point, the most salient features that are useful in discriminating between classes of item.

The third type of software system for vision is more like a conventional programming language, with the inclusion of vision operations. In this form of software some basic image processing routines have been packaged as macros or subroutines for use in a conventional language. In a typical Pascal-like system there would be routines to take pictures, find maximum dimensions of regions, compute the locations and features of various segments, etc. The advantage of this type of system is that a specific routine can be built to deal with particular checks during inspection tasks. A feature of inspection is that certain areas of the image are very critical while others are not at all important. The only way of knowing which areas matter is by utilizing information about the current task. Consequently, a new program must be written for each new inspection task. This type of programming language allows inspection routines to be produced relatively easily, at a fairly high level of task description, and without too much research into the design of different operators. Figure 3.21 shows such a task and a small segment of idealized code. The current trend is towards menu-driven, personal computer-based systems that eliminate the need to program in a language.

There are two generations of commercial vision systems. The first-generation systems emerged in the late 1970s and were based on binary image analysis. They used thresholding techniques to segment the image into blobs and were based on the belief that features in the segmented image corresponded to features in the real world. These systems are still used for applications where a silhouette or strong boundary feature can be used to process an image.

Second-generation vision systems can now be purchased and these are edge-based systems which process grey-scale images in order to analyse different light levels within the scene. These newer systems can locate a wider range of object features, other than just boundary-related variables. They are used in measurement and inspection tasks. The first generation systems were usually based on a turn-key design, whereas second-generation systems are more likely to be development systems.

The main problem with commercial vision packages is that they have to be general purpose in order to be applicable in many situations. This very requirement sometimes means that they are not suitable, or are over-complicated, for a particular task in hand. In robotics there are strong assumptions about the world that is being viewed — for example, industrial robot worlds are usually well structured and have strong expectations. Vision is not used in an exploratory sense but is used to confirm or measure or refine existing known data. This means that less sophisticated systems are quite useful in the industrial arena. Advanced techniques are not necessarily desirable for industrial robotics, and vision sensors are often required to act like smart sensors rather than intelligent entities in their own right. Once such a vision system has been programmed it will perform particular inspection tasks and return the results as though it were any other sensor. Its internal processing can be considered as a smart or pre-programmed activity.

Whichever commercial system we purchase, we are likely to use it for applications such as those listed earlier in Figure 3.1. If we consider the opportunities that such systems offer,

Artificial sight

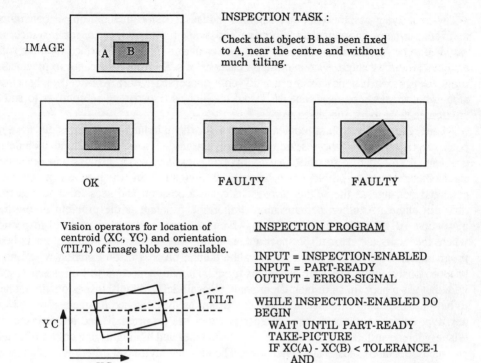

Figure 3.21   Vision system programming

we see that they can process images in a limited way, they can search for pictorial features, and they work particularly well if there are some distinct two-dimensional feature attributes that are to be recognized. In engineering and manufacturing, two-dimensional views are part of the language of component design and specification. If a component has a finite number of stable states and can be presented individually to a viewing system, then three-dimensional data can be decomposed into two-dimensions and successfully handled in recognition, inspection and measurement tasks.

The assumptions made in many robot vision systems are as follows:

- objects should not occlude one another
- ideally, objects should be presented one at a time
- the control of shadows and lighting should be strict
- the camera geometry should be known or fixed
- any relative movement between camera and object should be restrained and controlled.

The underlying assumptions that simplify industrial vision include: object constraints, restricted camera geometry, and lighting and scale factors. Objects, for instance, are assumed to be rigid, hard and usually have flat or uniformly curved surfaces. Unlike scene analysis, where we might encounter major changes of scale from landscapes to precincts to room interiors, in robotics we are usually only concerned with a work-cell or bench-top area. Another assumption is that the visual medium is transparent, in contrast to smoky environments or murky undersea work.

There are a number of hard problems which still exist in industrial situations that have not been solved outside of the research laboratory in any effective way. These include the problem of recognizing flexible objects involving textiles, wires, plastics and other non-rigid objects. Another problem is robot vehicle navigation, in which views of the world constantly change as the vehicle moves its camera position and sees fresh scenes from different angles. Another, perhaps more important, problem is the problem of *implicit* inspection. All of the inspection examples given so far refer to *explicit* cases of inspection, where the faults are defined or specified, in some way, at the time the system is being programmed. Implicit inspection refers to the human process of observation where novel but 'obviously' faulty items are classed as rejects. Human inspectors will often reject some item that has a poor finish or blemish or some other ill-defined fault. Inspectors do not need to be told all of the different faults that can occur, as they can see for themselves when a new type of fault, unique failure or other problem has occurred. These intrinsic characteristics include surface features such as dents, scratches and highlights, the colour of objects and their orientation, shape and reflectance. The ultimate high-level criterion is whether the object has been rendered unsuitable. But how is this idea to be programmed into the precise world of the computer? These problems remain very difficult research issues.

3.8 Future developments

There are various commercial possibilities for the third generation of vision systems. Given that objects can now be processed at the level of shapes in the image, we might expect that the third generation will offer full three-dimensional shape recognition and analysis. While there is considerable research progress in this direction, and some forms of shape detector may emerge, it seems that the expense and complexity of such systems will be inappropriate for industrial sensors.

However, depth information is more likely to become available as there are many techniques that can be used to recover scene depth. These range from stereo sensors using image feature correlation to structured light techniques where light source geometry is used to calculate distance. Other uses of structured light are probable, as this can be a cheap and accurate method for examining selected areas in a work space. With the advent of computer-controlled lasers and other electro-optics, the technology might shift from complex cameras towards dynamic light sources that actively interrogate the scene. This will lead to integrated packages consisting of sensors, illumination sources and correlation and control systems. Other specialized structured light systems are likely to be based on fibre optics and scanning systems for localized inspection and wear analysis applications.

While the interpretation of shape from image data is a more difficult problem than depth perception, there is an even harder task in motion detection. Image dynamics and motion

Artificial sight

properties are difficult because they require operations over time as well as space. This can only be achieved at present by operating over successive image frames (slow), or by using random access cameras (expensive). Another technique, known as optic flow, computes an output image consisting of an array of velocity vectors of all the smoothly moving image parts. While schemes for analysing observer and scene motion from optical flow images exist, it is not clear how such images can be efficiently utilized in real-time for the industrial context.

One of the most exciting new developments is the range of high speed and novel hardware architectures being developed for vision work. There are different approaches ranging from the implementation of fairly conventional vision operators and functions on VLSI chips, to the construction of full-scale novel architecture computers. Vision chips are not necessarily parallel devices in themselves, but can provide tremendous speed improvements over software algorithms by being used as components to build up parallel systems. Notice that local operator processing is ideally suited for parallel implementation as the mask logic can be carried out by a series of identical operators all working together. The resulting speed-up over serial methods will be approximately proportional to the number of parallel operators used. Such parallel vision hardware is under development at several research laboratories and advanced computer architecture for vision work is now a very active research field.

One of the most unconventional architectures for a parallel computer is the Connection Machine which was developed out of work at MIT. Although very expensive, the Connection Machine will be able to compute the results for millions of large operators in real time. This is due to its tens of thousands of processors organized in a richly connected network. The Connection Machine may become the major vehicle for testing full-scale computational theories of human vision.

Parallel architectures will also emerge in the industrial area. Machines, for example, like those based on the Clip series are especially designed and engineered for high-speed vision work. Such systems are now commercially available and will become common as advancing technology keeps hardware prices falling.

3.9 Summary

1 Imaging sensors can generate vast quantities of data — orders of magnitude higher than for other types of sensor. This factor alone makes imaging problems qualitatively different from those of other sensors.
2 Image data contains both spatial and temporal structure.
3 Image processing involves filtering out the relevant from a mass of redundant and irrelevant data.
4 Industrial vision systems must be fast, cheap, flexible and robust.
5 There exist well-tried techniques for recognizing two-dimensional objects. Three-dimensional objects are much harder to handle.
6 Pattern recognition methods produce classifications, usually from pre-defined structures. Image-understanding systems produce rich descriptions of images and interpreted objects.

7 Many global image-processing operations can be performed by local methods. Common operators are smoothing, edge detection, thresholding and noise removal.
8 Most industrial recognition and inspection systems operate by making decisions based on computed properties of two-dimensional shapes.
9 Commercial systems are designed as either turn-key systems, extended programming languages or development packages.

3.10 Further reading material

For a good text on a range of vision topics, see *Computer Vision* by D. H. Ballard and C. M. Brown, (Prentice-Hall, 1982).

There are several textbooks on basic image-processing algorithms and operations, e.g. *Digital Picture Processing*, by A. Rosenfeld and A. C. Kak, (Academic Press, 1982), and *Digital Image Processing*, by R. C. Gonzalez and P. Wintz, (Addison-Wesley, 1979).

Details of current industrial vision systems are best found in the latest manufacturers' catalogues.

For examples of vision technology used in industrial application see *Automated Visual Inspection*, edited by B. G. Batchelor, D. A. Hill and D. C. Hodgson, (North-Holland, 1986). A bibliography of automated inspection techniques is given by R. T. Chin in the journal *Pattern Recognition* (1982), volume **15**, number 4, pp. 343–357.

For some details of the massively parallel architectures that are being developed see *Parallel Computer Vision*, edited by L. Uhr, (Academic Press, 1987).

Chapter 4

The problem of perception

We find certain things about seeing puzzling, because we do not find the whole business of seeing puzzling enough

Ludwig Wittgenstein

In this chapter we continue our examination of computer vision in order to illustrate the difficulties of processing sensory data. In fact, these problems apply to most sensors, but they are easier to appreciate for computer vision and image processing.

4.1 Perception involves interpretation

We have described the subject of image understanding as the problem of trying to interpret images of three-dimensional scenes through a two-dimensional camera or sensor. This is a very difficult problem because for any particular two-dimensional image which is created as a projection of a three-dimensional scene there are an infinite number of scene interpretations that could have produced the image. Without knowing the depth of each individual illuminated area it is impossible to say where those components lie in the scene and so there may be many reasonable, but qualitatively different, interpretations of the objects that produced the image.

Figure 4.1 illustrates how an elliptic image shape can be produced by different scene arrangements. Although some of these may seem contrived or pathological, remember that, in engineering, orthogonal views are often preferred and so the end-on view of the elliptic spring is not unreasonable. Notice also that while a loop can usually be distinguished from a disc, on the basis of internal and external intensity differences, it is not easy to decide if it is elliptical or circular. Even the perception of a known ellipsoid can be mistaken; an oblique view can so foreshorten the image that the major axis can be seen as the minor one!

In human vision, we are not aware of this effect, mainly because we have two eyes and stereopsis and other visual processes provide the missing information, i.e. depth perception. But in single image sensing we have no way of dealing with this difficulty. The problem is even deeper than that, however, as it is not just depth information that is missing, but

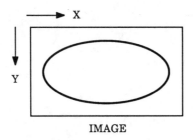

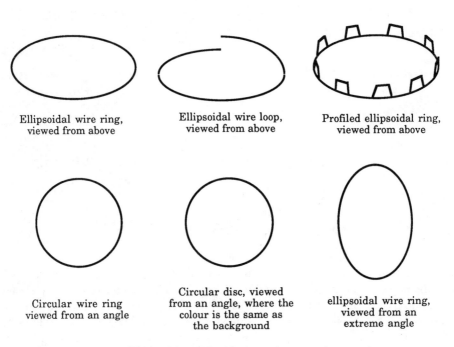

Figure 4.1 Multiple image interpretations

computers also lack an underlying semantics of what the scene is supposed to be or mean. For example, consider an image of a tree. Any given tree will be completely distinct and probably have details that are totally different from any other tree, and so there can be no such thing as an archetype tree with all the required and essential features. Furthermore, one interpretation of a tree, which is acceptable to one person, may not be satisfactory for another. Thus, there is no easy set of parameters that can be used. For example, the ratio of r to h, for the symbolic tree shown in Figure 4.2, might be used in a rule to distinguish a tree from a bush. But this is completely application-dependent: although it could be critical for one type of decision, it can have no general validity. There is no precise or fixed way of distinguishing all trees from all bushes. Also consider the variation between trees in summer, with maximum foliage, and in winter, when there are bare branches. There are also

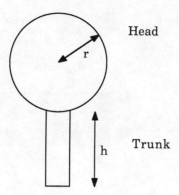

Figure 4.2   A symbolic (toy) tree

holly trees, oak trees, miniature trees and symbolic trees. Figure 4.3 indicates this wide variation. All of these have the essential characteristic of 'treeness' but it is very difficult to capture as a concept. There is no set of sufficient and necessary conditions for treeness. And yet, if we do not try to capture the idea of treeness, in some representation, we will not be able to ensure that our computer system has the ability to appreciate the variation that is involved in understanding this concept.

Figure 4.4 shows a simple example. The 'Wizzo Apple Sorter' is a device which will sort apples into large sizes and small sizes. Apples are poured in at the top and they roll out through two slots. Clearly, this is not a very difficult task, and some kind of mechanical

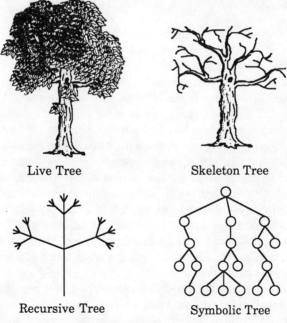

Figure 4.3   Kinds of trees

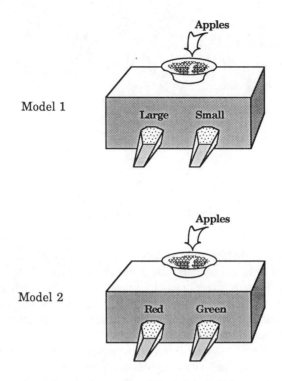

Figure 4.4   The 'Wizzo' Apple Sorter

screen with circular holes will be sufficient to achieve it. Such a simple unintelligent device is very efficient because: (a) it achieves the task required, (b) it is very fast — there are no computational overheads — and (c) it is infallible in the sense that large apples always come out of one hole and small ones out of the other hole. If we compare this with human sorters who are asked to segregate apples into large and small we would find that their error rates were much higher. However, such a device is not versatile: it cannot switch to another kind of task. If, for example, we now wish to sort the apples on the basis of their colour, so that red apples come out of one exit and green out of the other, then we have a much more difficult task. It is quite likely that human operators will be better at this, in terms of economic costs and speed, than any machine we can devise. The reasons for this are worth considering. First of all, 'large' and 'small' are well-defined and well-understood properties, whereas colour requires a complicated specification of a series of wavelengths to determine the acceptable shades of red or green. There will be difficulties with colours that are a mixture of the two, and also we have to define how much of the apple surface must be red before it can be classed as red. Altogether, this is a much more difficult requirement for a machine, although it has been done.

However, on reflection, perhaps the size parameter is also a less precise goal than we really thought. Although we can define a size test as being a circular hole through which an apply can fall, this does not really give us the three-dimensional size of the apple. The 'quality' of the size should also take into account the shape — whether there are lumps or

The problem of perception 57

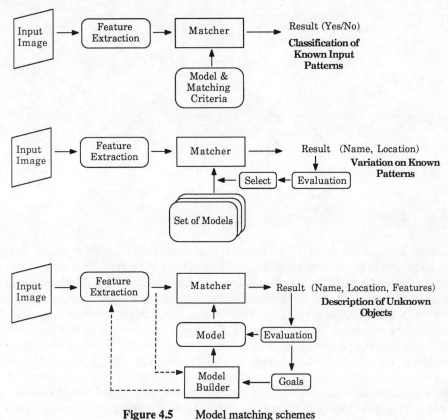

Figure 4.5 Model matching schemes

dents and so on. This leads us towards the requirement for implicit inspection. If there are some rotten apples or damaged apples, or apples with funny features or blemishes, these would not be detected either, whereas in a more sophisticated device that looked at colour such extra tests could be more easily incorporated. Also, it would be useful to measure trends. Human operators would notice if the apple quality was deteriorating, if the apples change from 'Granny Smith' to 'Macintosh', or if some other parameter changes. These are implicit features. Then again, there may be new classes of apple, a special apple might occur which was softer, longer, and didn't roll very well. In time we might decide that these objects were sufficiently different to merit the creation of a new concept, i.e. a pear. The words 'pear-shaped' capture a general but precise definition of this particular shape. This simple illustration shows that there must be more than just the matching of parameters and straightforward tests of equality. Something about the *quality* of the parameters matters, not just their threshold level.

Figure 4.5 illustrates the range of sophistication that can be used to increase the power of matching and model processing. In the first level, image features are extracted and simply compared against a prototype model to see if the required features are there in sufficient evidence. Such a system is acceptable for pre-determined inputs and pre-determined classes of output. The next level shows how the matcher can have access to a set of models and an

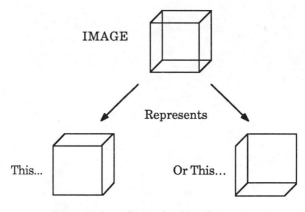

Figure 4.6   Competing hypotheses

evaluation, based on the results of the matching, can then select different models if it didn't find a good match. In other words, if it doesn't appear to have a very good apple, it can try to match it against pear and see if this produces a better result. Such a system is useful for variations on known classes of things. Finally, the third level shows a more complex system, with more feedback paths, in which a goal is specified to a model builder that allows classes of models to be generated. This type of system is useful for building descriptions of unknown items until sufficient evidence can be amassed to produce an output result. We can see this last example as a move away from analysis towards synthesis. The system now builds a hypothesis for the events. If the hypothesis matches the external world, then we believe we have a good hypothesis and have solved the problem — we recognize. This is the process of perception. We only *receive* an image, but we *perceive* an object. In other words, some processing must be applied to the raw data in order to make sense of the sensory variables. Images themselves are just arrays of numbers but the meaning of the image is much more than that, it is an interpretation (or a perception) of an external situation. With such hypothesis systems we have far more power. We can now perform predictions using the model, we can estimate and plan and we have much more flexibility. A danger is that erroneous hypotheses may occur. We may choose the wrong model and base our future actions on that. This is analogous with human visual illusions. In such cases, there is not enough information to resolve ambiguities of interpretation and illusions can occur. Many famous illusions have these properties, for example, the reversal of certain line diagrams, as in the Necker cube shown in Figure 4.6. Usually there is a great deal of redundant information in visual scenes and these effects are not noticed. We rarely even glimpse the complicated processes that must support perception in our everyday life.

4.2 The analysis of three-dimensional scenes

We will examine scene analysis as an illustration of the massive difficulties that are faced by image understanding research. Figure 4.7 shows a schematic diagram of the way in which an image is processed to gain an understanding of the objects which are represented in the scene. First of all, local operators are used to segment the image, either into line

The problem of perception 59

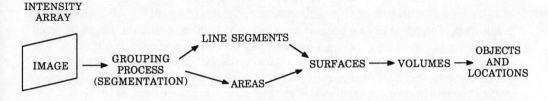

Figure 4.7   Scene analysis

segments or regions, or perhaps both in parallel. Then the line segments or areas are grouped to form surfaces. The surfaces in the world represent the boundaries of volumes. Notice that these surfaces will change as we move viewpoint, hence, these are viewer-centred representations. The next stage is to construct interpretations or representations of objects from the surface structures, thus forming volumes. Finally, the volumes are identified as objects and recorded as entities that occupy space and have physical attributes. Objects are not viewpoint-sensitive but are independent of the way in which the picture was taken, and are therefore object-centred representations. One method of defining volumes is to use the generalized cone (or cylinder) representation. This consists of a space curve known as a 'spine', along which a cross-sectional area is swept while being modulated by a sweeping function. By using all kinds of cross-section and sweeping function we find that many every-day objects can be represented by this technique (see Figure 4.8).

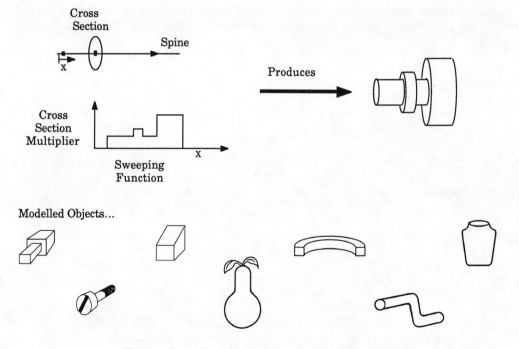

Figure 4.8   Generalized cone (or cylinder) representation

There are other methods of representing three-dimensional volumes; CAD systems often employ kits of standard blocks, prisms and cylinders to build up composite shapes by union and intersection operations. Figure 4.9 shows this alternative representation concept. The generalized cone method is particularly good for 'prismatic' objects: extruded, machined or turned components. It can be awkward for representing castings, moldings or fabricated items. Generalized cones can also model 'natural' objects to some extent, as in Figure 4.8.

The trouble with real-world scenes is that they are so complex and so variable that they become unmanageable for most unconstrained real-life applications. This has led to a history in which underlying assumptions have been made about various features so as to introduce constraints and limitations on the image. In Figure 4.10, we see the range of constraints, many of which have been used to help make the problems much more tractable. For example, we might restrict our objects by selecting particular textures and finishes. We might choose very smooth flat objects, their reflectance could be designed to be a medium tone with little variation, and the lighting and the environment could be controlled, so that shadows are very limited, if at all, in existence. In addition, structured light or special camera positions can be used to control the picture-taking process. The objects may be selected from a few restricted classes so that the scenes could consist of only a set of blocks, a group of engineering components or perhaps a certain class of houses of a particular sort or style. Any constraint on the objects helps to solve the problem. The scene complexity can also be controlled: we may perhaps look at only one object at a time, or some multiple objects which do not overlap or occlude one another, or at the extreme, we might use mixed objects of different textures, shapes, positions and locations in space. The problem here is that whatever constraints we adopt we may well find some method of solving the problem to produce reasonable scene interpretations, but if we ever decide to lift a constraint and go for a slightly harder problem, then we have to re-run the whole research program or design

OBJECT:

Is composed of 5 primitives :

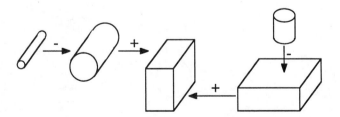

Figure 4.9   Solid modelling by composition of volumes

The problem of perception 61

```
INTERNAL SYSTEM CONSTRAINTS

Hardware processing capability      - speed
                                    - storage limits

Software functionality              - computational complexity
                                    - algorithm quality

Sensory attention controls          - selectivity
                                    - expectations

EXTERNAL CONSTRAINTS

Sensor characteristics              - bandwidth
                                    - digitization precision
                                    - sensitivity

Sensed object properties            - shape
                                    - texture
                                    - surface finish

Scene configuration                 - shadows
    lighting geometry               - perspective
    camera geometry                 - special sources
    structured light                - coordinated cameras

Environmental variables             - movement of objects
                                    - movement of sensors
                                    - stability of scene
```

Figure 4.10   Constraints in visual sensing

process in order to try to find the parameters of the new situation. There is no reason to believe that a piecemeal lifting of constraints is in any way feasible and in fact there is some evidence that it will not lead to a monotonic increase in understanding.

4.3 Blocks worlds

For the type of scenes which interest us in industrial environments, we frequently have machine-made components with fairly formalized geometric shapes. A considerable amount of 'blocks world' research was done in scene analysis in which very straightforward hexagons, wedges, cubes and other simple geometric shapes were used for recognition. There are two main paradigms in blocks world scene analysis: *model-driven* and *picture-driven* methods.

In the *model-driven approach* the system maintains internal models of, say, an abstract cube. In this type of system when a new image is received it will be matched against transformations of the internal cube until a suitable transformation can be found that matches the image. If this is found, then the system assumes that the image represents a pictorial version of a cube. If the cube fails to match, then another model may be taken from the library, such as a wedge shape, and this may then be transformed to see if it can be made to match the image. In practice, guiding information will be used to draw from the image the most likely models to test first. The difficulty with this system is that if the shape is a new form, and no model has been included in the library, then the system may well fail or, at least, produce very strange interpretations. On the other hand if a known class of items is to be matched, then the system can go to great lengths to try different transformations until it finds the correct viewpoint to match an object.

The other method of approaching the problem is the *picture-driven* or *data-driven approach*. In this case the picture is scanned until various cues are found which indicate certain structural properties that occur in three-dimensional objects. If we are working in the blocks world, then we will find that the corners of blocks have certain pictorial characteristics. For example, a fork shape is often seen at the corner of a block, an arrow shape is often seen where one edge of a block is nearer to the camera than the other two edges, and so on. A number of researchers developed these ideas, culminating in the work of Waltz, who developed a very comprehensive geometric database for all the different ways in which a corner or an edge could be interpreted. Waltz's technique was then to take an image and look up the different interpretations in the database and associate these with the vertex points where the lines joined in the image. Because the representation of any given edge must be the same at either end of the line, conflicting labels at different ends can be cancelled out. By this process, the Waltz algorithm was able to wander around the diagram crossing out inconsistent representations until it was left with only one or two different labellings for the image. This proved surprisingly effective for the limited class of blocks world scenes. The label interpretation process is a constraint propagation method. This is very similar to relaxation programs which process various field equations until consistent values have been accepted throughout the system. The surprising thing about Waltz's work was that despite the large number of possible labellings that could be attached to any given line junctions, only a few of them are actually realizable in practice. For example, in one case, there were 57 variations for labelling a line, so a Y-junction with three lines had 57^3 combinations of labels, that is, roughly 185 000, but in practice *only* 86 of these are physically realizable. Because of this discovered reduction from the theoretical to the possible, powerful convergence methods become valuable. As a label must be the same at either end of a line, many of the large number of combinations that might fit a picture are cancelled out due to the constraints that propagate through the picture lines from all the other junctions. (The secret of Waltz's performance results was in the once-only pre-compilation of the geometric labelling possibilities. The picture processing could then use rapid look-up tables and elimination methods.) This idea of constraint propagation to eliminate possibilities has found other applications in AI and is sometimes referred to as the *Waltz filtering effect*.

Although Waltz managed to get impressive results with labelling techniques, even for the edges of shadows, there is no future in trying to develop larger sets of labels to cover more complex objects in the scene. Not only will the number of different edge types become very large but many of them may have little significance, e.g. reflectance patches on objects.

Also, curved lines can have different labels at their ends and so most of the value of this labelling technique disappears.

We could cite other examples of image-understanding research but this is such a vast field in its own right that it is not profitable to cover it all here. Suffice it to say that success has also been achieved in other specialized areas — in the interpretation of satellite photographs, in the processing of images of semi-natural scenes such as houses, and in surveillance (for example, car number-plate recognition). But the underlying problem of *understanding* what is in the scene remains an application-driven problem. If we do not have any way of informing the system or describing the important semantics that underlie the scene, we will not be able to do the processing that is required for recognition, inspection, or interpretation tasks.

This message we will see again in expert-system research, but it seems to be one of the major results of AI work: the use of application information provides major guiding principles in reducing the complexity of the processing problem. In other words, if we wish to use vision in an industrial task, we must be very careful to define exactly the parameters of the task, exactly what is to be measured or inspected and how it is to be handled, in order that we can implement a reasonable system with high reliability and fairly low cost. Of course, this is also true if we wish to implement the ideas of implicit inspection, as described in Section 3.7; we must find some way of defining the implicit parameters. This does not mean that we have to enumerate them all, but we do have to include some model of acceptability so that the system is able to manipulate this model and determine when the parameters or the boundaries have been over-reached. It is quite likely that this can be done, but it requires very sophisticated representations of the objects, the task and the processing requirements in order to do it. We should not expect such systems to be readily available off the shelf, at a low price, for a few years yet.

In the meantime, we can use imaging sensors to provide us with quite sophisticated recognition features. We can use them to locate and measure the position of objects, and we can use them in limited inspection roles, provided that the parameters are well determined. It is probably best to consider industrial vision systems as smart self-contained sensors with well-defined roles, rather than as general vision facilities.

It seems that the simple one-at-a-time process diagram shown in Figure 4.7 is, in fact, far too simplistic for general vision work. It is too easy to suppose that one layer of processing will output results which feed into the next, and so on, in a chain of increasing interpretation. There may be quite complex interactions between the processes, for example, an image line may be missing and at a later stage, when reconstructing an object, we may wish to interrogate the line finder to see if it could reasonably agree with the suggestion that a line should be there. Such systems involving many messages and complex communication between their modules have been experimented with but are quite difficult to handle in a formal manner and rather hard to evaluate out of the research laboratory.

4.4 Current research directions

The emphasis on vision has now moved away from the blocks world towards a more general model of vision which is compatible with what is known of human visual systems. Researchers are now working on general theories of vision based on computational models.

They are trying to work from general abilities to specific applications; in contrast to the blocks world where they attempted to abstract general principles out of particular case studies. They are also trying to discover general visual-processing constraints that affect the whole visual system, rather than constraints drawn from applications. Finally, there is an emphasis on mathematical models, in which the phenomena found in smoothing and gradient methods are unified in one treatment, thus giving a better understanding of the options available. The sort of constraints that are now being examined are (1) look for the most conservative solution, e.g. use the simplest space curve to fit the shape being examined, and (2) only points of rapid change are considered interesting. Thus, areas of activity and novelty become important in the scene.

The search has shifted back to low-level processing which is thought to be fundamental for all other visual processes. It is thought that no knowledge of high-level models is contained in the lowest level substate of vision. But there is some *general* visual knowledge stored there, that is, good representations and useful attributes of the image. Research is now trying to discover general scene analysis knowledge, using ideas based on properties of regions and contours, rather than using specific domain-based knowledge. Local processing is still very much in evidence using gradient detectors and convolution methods.

There are also many results on the generation of shape data from various forms of *image information*. Shape-from-stereo is a process in which the shape and depth of surfaces from the viewing point are computed, either by analysing slight disparities between two images based on raw image correlation, or by using correlation between edges and edge-matching techniques in the image. These two methods have different characteristics. Image correlation gives lower accuracy but is good for texture and smooth areas, whereas edge-matching has the opposite characteristics. Other methods of analysing shape include *shading*, where the gradual fading away of light intensity on surfaces gives clues to the curvature of those surfaces. Given that lighting positions are known and that the surface has a uniform reflectance characteristic, then shape can be computed from image shading patterns. Another shape-finding technique is to measure any *texture* in the image. This is the well-known perspective effect where more distant objects with a regular texture appear to have smaller patterning than near objects. Yet another method is *shape-from-contour* analysis in which the perceived slant and tilt of the surface is used to calculate the probability of the surface orientation. Finally, another method is the use of *motion* to detect shape. Objects that are nearer to the observer appear to move more rapidly and therefore changing scenes can be perceived in terms of the flow of dynamic shape information in the scene.In this method the image velocity directions give clues to the movement of shapes in the scene.

It is interesting that local operator processing is still useful for three-dimensional scene analysis and in fact nearly all global processes can be implemented in some form of local processing. Clearly, this is true for the animal visual system where large numbers of parallel neurons are arranged in layers of processing functions. Current research work on vision tries to emulate some of this structure.

The basic assumptions for most objects are that the scene shows a not too-oblique view of a smooth finite-sized surface. Accidental alignments of objects, shadows and boundaries are quite rare and these are normally ignored. The basic image processing substrate is considered to be an intensity change map, sometimes called the *primal sketch*. From this representation, edges can be detected and grouped into line segments and then the lines are interpreted as boundaries. This produces a surface representation which is viewer-centred. The next stage is to extract volumetric information in order to get a non-viewer-centred representation. This is perhaps the most difficult step in vision. Various methods are possible including the intersection of archetypal blocks or the use of generalized cones.

4.5 The problem of understanding

It should be clear from this discussion of the difficulties of processing three-dimensional information from a two-dimensional image that vision, in general, is a highly sophisticated and complex facility. We would not expect the full power of an artificial general-purpose vision system to become available for industrial robotics tasks for many years. But, even if such systems were available, it may be unwise to pursue this for industrial needs. Industrial vision has different requirements from general-purpose vision systems and we can capitalize on this in terms of less sophisticated but highly effective systems. Figure 4.10 listed some of the important constraints in vision sensing that can be artificially, accidentally or deliberately introduced into a vision task. We can see these in all forms, from internal constraints to external adjustments to the physical world. It is the very existence of these constraints that make vision problems tractable in the industrial domain.

4.6 Summary

1 Perception is much more than parameter estimation; it involves the interpretation of complex data.
2 Industrial vision has different objectives, methods and constraints from those of 'general' vision.
3 Implicit inspection methods are very desirable for industrial applications, but are difficult to achieve because they involve the problems of perception.
4 Constraints, of any kind, will be helpful in reducing the difficulty of a vision problem.
5 The results of over-simplified 'blocks world' experiments are often not relevant for industrial vision.
6 Vision research is developing techniques for recovering shape information from images by using: shading, texture, contour, motion and stereo.
7 The use of application information provides major benefits by reducing the complexity of visual processing problems.

4.7 Further reading material

The vision research issues discussed in this chapter are reviewed by M. Brady in *Computing Surveys* (1982), volume **14**, number 1, pp. 3–71. Also, the journal *Artificial Intelligence* has a relevant special issue on computer vision (volume **17**, numbers 1–3; 1981). The IEEE *Transactions on Pattern Analysis and Machine Intelligence* are a source of research results.

The book *Computer Vision* by D. H. Ballard and C. M. Brown (Prentice-Hall, 1982) can be highly recommended for insight into these topics.

An interesting discussion about 'Expert vision systems' is in the useful journal *Computer Vision, Graphics, and Image Processing* (1986), volume **34**, pp. 99–117.

For Marr's seminal work on computational models of human visual processing, see *Vision* by D. Marr (Freeman, 1982).

Chapter 5

Building a knowledge base

What is all knowledge too but recorded experience, and a product of history; of which, therefore, reasoning and belief, no less than action and passion, are essential materials?

Thomas Carlyle

5.1 Introduction

In the previous chapters we looked at the difficulties associated with sensory processing and now we turn our attention to the design and construction of knowledge bases which form the heart of intelligent systems. Such knowledge bases contain information derived from many sources, including sensors, and are used to support reasoning and planning processes. A knowledge base is a store of information but should not be confused with a database. Knowledge bases record data but are also able to manipulate, refine, modify and create data. Consequently, they are not just passive stores of facts but involve processing and self-modification.

Another distinguishing feature of intelligent knowledge-based systems is their use of 'heuristic' information. Heuristics are unguaranteed 'rules of thumb' that offer advice or guidance to control a process or aid a decision. Heuristic methods do not give guaranteed results, unlike algorithms, but represent experience and other empirical knowledge when more rigorous or complete data are not available.

There are many aspects to the design and construction of knowledge bases and Figure 5.1 shows one way of viewing the topic. The first issues to be considered are the ways in which data can be structured and interrelated. This is a major research area in AI. Another central feature involves the inherent special characteristics of the data itself; uncertainty, inconsistency and general lack of integrity are serious problems. Finally, there are also considerations regarding the general operation and functioning of knowledge bases. For example, a requirement for good explanation facilities might lead to additional structures being incorporated, so that operational knowledge (i.e. meta-knowledge) might be stored and utilized. We will examine each of these three areas in turn.

A good question to ask is, why do we need a knowledge base? Why do we not simply use an ordinary computer program to control our processing systems? The answer is that we

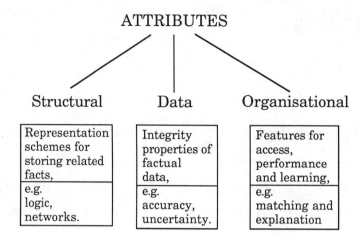

Figure 5.1 Attributes of knowledge based systems

need more than a conventional program can provide; we need intelligence. For example, we want our system to be able to reason about things which it already knows, to deduce answers to problems, to respond to new situations and learn about new facts – and all this without too much human intervention. It must also be able to report to us and explain its actions in a suitable language. It should not communicate in obscure programming or computing terms but should use every-day knowledge and terminology in order to talk to the 'client' about his or her application. In order to facilitate this and produce a system that can be accepted as 'intelligent', all of the attributes shown in Figure 5.1 are essential considerations, and most of them will have to be resolved in some detail during any implementation.

There are two main difficulties that we must deal with. The first is the capture, collection or compilation of known facts or data about our particular task, in other words, the acquisition of available knowledge. This is the *knowledge acquisition (KA) problem*. Secondly, we need a method for representing this knowledge in the most appropriate form. This is the *knowledge representation (KR) problem*.

The first problem is really one of coaxing the knowledge out of expert personnel, or from whatever other sources are available in the application domain, and making it as explicit as possible so that it may be coded into a computer system. This is a difficult problem area and there are few real guidelines. We do not yet have a good theory of KA and current techniques such as interviewing and protocol analysis are more of an art than a science. The most common way of capturing knowledge is by interaction with an expert in the field, but there are other ways, for example, by exploring an environment, by the use of self diagnosis or discovery, or by other forms of learning. Detailed studies of experts such as master chess players or senior medical practitioners tend to emphasize how much they don't understand: not about their subject, but about how their own expertise works! In engineering, the experts also have backgrounds based on years of training and development, consequently their flair and intuitive insights are equally difficult to define and yet equally important. A key problem in knowledge acquisition is deciding which facts are the most relevant and how they can be evaluated in the context of the problem domain. Because of these problems most systems draw their knowledge-base facts from the application area in a rather *ad hoc* manner. There is a considerable body of work now going on in KA, in particular, in the study of tutoring systems and human learning research; however, we are not likely to benefit from these areas for many years. In our engineering domain, our only guideline must be to capture, explicate and record all possible data covering our problem area. Fortunately, engineering has a strong advantage over other fields in that a great deal of proven and practical information has been recorded in the literature. There are many well-documented methods and case studies, and much application data are now becoming available on computer media.

The second problem, how to represent knowledge, is also a major research area in artificial intelligence. The KR problem addresses representation issues for all of the different aspects shown in Figure 5.1. This is a crucial topic because the choice of a bad knowledge representation method will nearly always produce a dismal failure. A 'good' or even 'appropriate' method is likely to be more successful — if only we knew how to estimate such factors — but does not guarantee success. So it is essential that a suitable representation technique is found for the nature of the task that is being handled. We need a good *notation* to express our knowledge and some of the requirements are as follows:

(a) The notation is appropriate. In other words, it is effective for the task that we have.
(b) The notation must be sufficiently precise that it can specify what is desired clearly and without ambiguity.
(c) The notation must be flexible in the sense that it can cover a range of different types of knowledge.
(d) Manipulation facilities must be available. This will allow facts to be created, refined and reasoned with.

In this chapter, we will examine different types of knowledge representation and the closely related topics of data integrity and knowledge-base organization.

5.2 Knowledge representation schemes

First consider a scenario for an example application. Assume that we have a small heat-treatment plant in a factory where a robotic device loads components on to a tray and then places the tray into an oven (see Figure 5.2). This is only a small part of the factory and it will be connected to other manufacturing machinery including the factory computers which manage and schedule the various workloads. In considering this scenario we see that the heat-treatment processing plant will have many activities and will be required to handle various types of intelligent activity. For example, it may have different orders sent to it at any time, it might have to deal with error conditions and failures, and it will have to report back to the other controlling systems, on events that happened and when and why it did various things. It will also have human operators who, from time to time, ask questions, make operating adjustments, or change the system's role by reconfiguring the plant. All these things could be handled by an intelligent system. This local intelligence approach will also minimize the amount of centralized programming effort that has to be invested each time a tiny change is required.

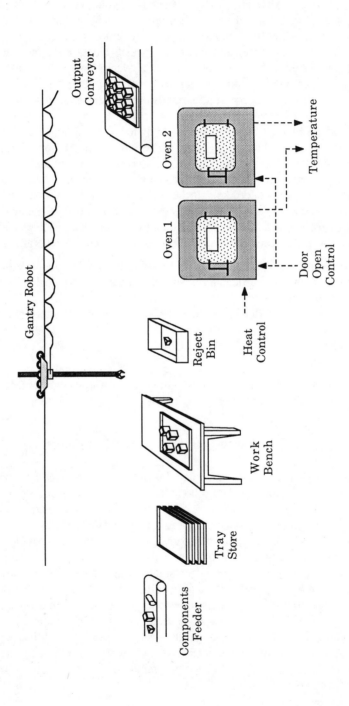

Figure 5.2 Robotic heat treatment plant

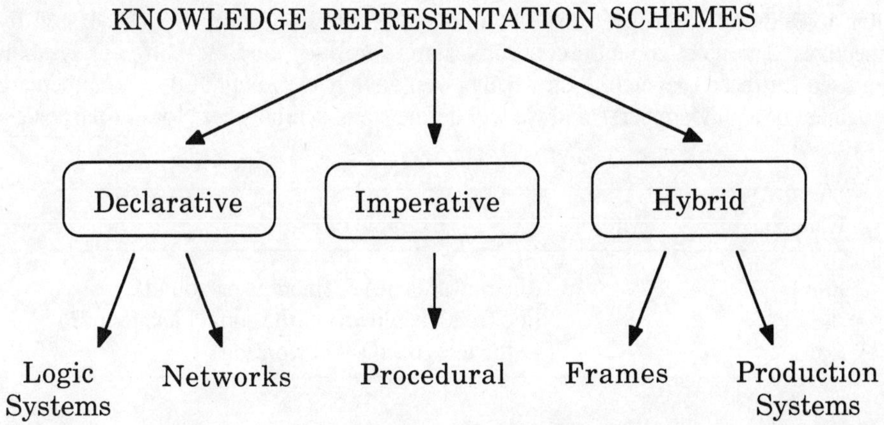

Building a knowledge base 71

KNOWLEDGE REPRESENTATION SCHEMES

```
         Declarative        Imperative          Hybrid

       /         \              |            /          \
  Logic         Networks    Procedural    Frames      Production
  Systems                                             Systems
```

Figure 5.3   Knowledge representation schemes

We will now look at different methods of knowledge representation, bearing in mind this scenario, and discuss the relative merits of the different techniques. Figure 5.3 shows a breakdown, in very crude terms, of the different methods of knowledge representation. There are two main categories of representation: *declarative knowledge* and *imperative knowledge*. Declarative knowledge refers to passive, factual statements whereas imperative knowledge is more assertive, being based on commands and actions. In other words, imperative knowledge is action-centered and declarative knowledge is data-centered. There are two main types of declarative knowledge representation — *mathematical logic* and *network methods*. There are also two good examples of *mixed techniques* using ideas from both declarative and imperative knowledge representation schemes. This makes five techniques in all and we will now examine each of these in turn.

5.2.1 Logic-based representation methods

It is worthwhile to outline the basis of mathematical logic here as, besides being a representation scheme, it is a useful notation for discussing many AI problems and issues in knowledge engineering. We will start this section by looking at the essential elements of mathematical logic and then show how these are used in computer representation schemes.

5.2.1.1 The propositional calculus
A proposition is a statement that is either true or false, e.g.

(television sets require electricity)

This is a proposition which happens to be true. Brackets are used here to delimit different propositions. Statements like this can be made up for any facts that need to be represented. Clearly, they will be very simple if they are limited to being either true or false and we use connectives to join them together and produce more interesting structures. Sentences can be built up from propositions and connectives and these sentences will have a truth value which depends on the value of the component parts and the operations defined by the connectives. The most common connectives are: AND, NOT, and OR. Different symbols are often used for these (depending on whether you have been brought up as a mathematician, an engineer or a philosopher!), and we will define some symbols here for our purpose:

Connective	Symbol	Example of use
AND	&	(the power is on) & (there is no sound)
OR	v	(the fuse has blown) v (the supply has failed)
NOT	~	~ (the television set is working)

We can now find the truth value of any sentence in the propositional calculus by assigning truth values, using symbols like 't' or 'f', to the individual propositions and then drawing out a truth table to find how the connectives produce truth results. Simple cases such as AND and OR are well known, but care should always be taken that they are not confused with the everyday usage of these words, which is often ambiguous. AND is true *only* if both of the two propositions connected by AND are also true. It is important to realize that logic is rather like a kind of algebra and is not concerned with the actual propositions themselves or whether they have any real meaning. So we must not think of any underlying significance or semantics of the propositions when we are performing logical operations. It is up to the system designer to make sure that the propositions are well chosen for the purposes of the application. Consequently, serious problems are likely to arise if meaningless propositions are introduced into knowledge bases as the reasoning system might corrupt the system by introducing many spurious results.

Another important connective is implication:

(there is a picture) $\rightarrow$ (the power is on)

Implication means the first statement *implies* the second. Thus if the first statement is true then the second statement will also be true if implication holds. We can use statement letters instead of writing out particular propositions, e.g. $A \rightarrow B$. Note that $A \rightarrow B$ does *not* indicate that $B \rightarrow A$ is true. Just because a television picture implies the existence of power, there is no reason to suppose that a supply of power implies a picture will appear. There is no causal connection between A and B. In fact, there are no causal relations at all in logic. Thus, we must beware of our intuitions when using implication. Let's look at its truth table:

A	B	$A \rightarrow B$
f	f	t
f	t	t
t	f	f
t	t	t

Building a knowledge base

We see that if A is false then B can be any value without contradicting the idea that A *implies* B. So A = true, B = false is the *only* case in which implication is false (i.e. it doesn't hold). Other symbols are used for implication including the word IF. In this case, we read implication as IF A THEN B, but as this can be confused with computer programming semantics we will avoid this notation. We will find implication very important later on.

All possible propositional statements can be written using only a small set of connectives. For example, using only NOT and AND, or using only NOT and *implies*, is sufficient to write any propositional statement. (In fact, only one connective can be used, as in the case for NAND (alternative denial) or NOR (joint denial) because these can produce NOT and AND, and NOT and OR respectively.) Such universal sets of connectives can also be valuable in knowledge bases because it means that only a few types of operation are needed and this can simplify and unify the processing operations. As an example of this notice how A → B can be written using only NOT and OR, as (~A v B).

5.2.1.2 The Predicate Calculus
We soon find that we would like variables in our propositions so that we can make statements like:

$(x$ has-plug$) \rightarrow (x$ uses-electricity$)$

In this case x represents a class of things so that the statement reads 'something that has a plug uses electricity'. Of course, this implication might not be true and we might not be able to find a value that fits x, but the structure is a much more useful arrangement than the previous fixed sentences. Notice that hyphens have now been introduced to group together all the words that relate to one concept or logical item. We will now use this notation throughout (if we were to leave gaps between words as in the first examples, each word would stand for a separate logical item, which is not usually intended). Now let's consider some notation. The component statements used above are called *formulas* and they each have a single predicate. In this case the predicates are: uses-electricity and has-plug, and they each have one argument, x. A predicate is a kind of logical function that evaluates to true or false depending upon the truth values of the arguments. Normally a slightly different notation called *prefix notation* is used in which the predicate is placed first. We could re-write the second formula as follows:

(uses-electricity x)   as in LISP

or

uses-electricity(x)   as in PROLOG *

We will use this prefix notation in future. This type of notation is more flexible, because we always know that the first item is the predicate and any following items are arguments. For example:

(built-from house bricks tiles wood cement)

---

* LISP and PROLOG are AI programming languages frequently used for building intelligent knowledge-based reasoning systems.

where the predicate, 'built-from', is followed by five arguments. Clearly, the order of the arguments is important. In the above example, the first argument, 'house', has a special relation to the other arguments. Each predicate has its own fixed number of arguments and its own requirements for the range of those arguments. Sometimes commas are used as argument separators and other variations of notation will be found in text books.

Predicates simply define a logical association between the arguments and are either true or false, so one predicate:

(made-of bolt brass)

might be true while another, with the same arguments,

(looks-like bolt brass)

could be false. In general, the truth of any predicate like,

(made-of x y)

will depend upon the values for x and y.

5.2.1.3 Quantifiers
In statements like:

(metal x) & (red x) → (hot x)

we need to define the range of x for this to be true. It is possible that lots of values for x will fit the statement, or only a few, or even none. There are two kinds of quantifier. First, the *universal quantifier*, $\forall$, which states that the range covered is *all* values of the variable. In other words, the statement:

$\forall x$ (metal x) & (red x) → (hot x)

means *everything* that is metal and red is hot.

Secondly, the *existential quantifier*, $\exists$, which says the range includes at least *one* value for which it is true. So the statement:

$\exists x$ (metal x) & (red x)

means there is *something* that is metal and red.
Another example:

$\forall x$ (instance x bolt) → (made-of x steel)

If this is true then all bolts are made of steel. So, although there may be other things made of steel, there are no bolts not made of steel.

Building a knowledge base

We have used constants and variables as arguments. Functions can be used to produce *values for arguments*, for example:

(size-of bolt)

This looks like a predicate but in fact gives values computed from the evaluation of the function, size-of, applied to the argument, bolt. So functions can be used in expressions, e.g.

(instance x bolt) & (large (size-of x)) → (heavy x)

These three — constants, variables and functions — are called 'terms'. So, in our system of logic, we have: *terms* (i.e. constants, variables, functions), *predicates*, *quantifiers* and *connectives*. The language *formulae* are built from predicates applied to terms, using quantifiers and building compound formulae with connectives. Such a system of logic is called a *first-order predicate calculus*. There are higher-order logics which allow variables to stand for predicates and there are different types of logic such as many valued logics, fuzzy logics, modal logics and probability logics. These all have different properties and have all been used for knowledge representation work, but by far the most well-known system is the first-order predicate calculus. Predicate calculus is the foundation stone and the other logics can be considered as variants or extensions that attempt to deal with further complexity.

We can easily build a knowledge base of facts. We simply store factual items as propositions, like:

(density steel 9.0)
(density wood 0.5)
(conductivity copper high)

However, a knowledge base requires more than this. We need facilities for manipulating data, refining facts and generating new knowledge from old. Logic has a very well-defined framework for this. There have been many research efforts directed towards logic tools for knowledge-base manipulation and general AI problem-solving. Theorem provers are the classic example here. They are well developed and have a long history in AI. The concept of generating new facts from old is known as inference. Figure 5.4 shows how, from a given

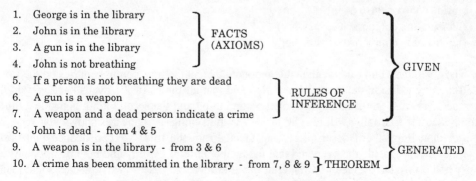

Figure 5.4 Producing new facts by inference

set of facts (known as *axioms*) and a few rules of inference, a number of new statements can be produced. These statements usually follow one another in logical sequence until a desired statement, known as a *theorem*, is reached. The program that generates the new inferences is known as a *theorem prover* and the steps generated by the application of the rules of inference are the *proof steps*. Theorem provers use the inference rules to attempt to find a proof for any given theorem. A great deal of research has gone on in automatic theorem proving and there are now quite sophisticated theorem-proving packages available. Most of the applications of these theorem provers have been in mathematics or in other rather abstract domains. There are a few examples of them being used for planning routes for mobile robot systems and in related areas, but they tend to be very expensive in computational resources and rather impractical for most day-to-day applications.

There are three types of inference which are of interest in AI. The first one, *deduction*, produces logically correct results as the deduced facts are legally derivable from the given axioms. The two other methods, *abduction* and *induction*, are not legal in the sense that the derived results may, in some cases, be unfounded. These are the types of inference methods used in expert systems and in learning systems respectively, and are discussed in later chapters.

Deduction is attractive because the results can be relied upon to be logically sound. There are various different deductive inference rules; perhaps the most well-known one is *modus ponens* which is stated as follows. Given that the implication a implies b is true and that a itself is true, then we may infer that b is true. There are several other inference rules but modus ponens is quite sufficient to illustrate the general idea. Consider the following example:

Axioms (given facts)
   (on component-a tray)   fact
   (on component-b tray)   fact
   (inside tray oven)   fact
   $\forall x$ (inside x oven) $\rightarrow$ (temperature x very-hot)   rule
   $\forall x\, y\, z$ (on x y) & (inside y z) $\rightarrow$ (inside x z)   rule

Inferred facts
   (temperature tray very-hot)
   (inside component-a oven)
   (inside component-b oven)
   (temperature component-a very-hot)
   (temperature component-b very-hot)

The given facts and rules state that component a is on the tray, component b is on the tray, and the tray is inside the oven. We are also told that all things inside ovens have very high temperatures and that if anything is on a second thing and the second thing is inside a third thing then the first thing will also be inside the third thing. Perhaps this shows that logic is better than English for describing some facts! From the rules we can deduce that both components are inside the oven and that the temperature of the tray and both components is very hot. So we have added new material to our knowledge base — we now have eight facts instead of the original three. This usage of the inference rules is known as *forward*

Building a knowledge base

chaining because the new facts are generated as forward consequences of the rules. Forward chaining involves pattern matching followed by the assertion of logical consequences.

If the inferred facts are inserted into the knowledge base as they are generated, then immediate answers will be available for later queries. If we now query the knowledge base by asking a question like:

(temperature component-*a* very-hot) ?

we immediately get:

true

as this fact is stored directly in the knowledge base.

An alternative technique, known as *backward chaining*, is often employed when the desired facts or consequences are not known at the time of query. In this case, the inference rules can be used in reverse to produce sub-goals that might satisfy the query. For example, the query:

(temperature tray very-hot) ?

cannot be found in the original facts. However, the right-hand side of the first rule will match this request if 'tray' is taken as a value for the variable x. The rule states that this (goal) will be true if the sub-goal:

(inside tray oven) ?

is true. This sub-goal is found to be true, as asserted in the third axiom, and so the query is satisfied.

Hence, any query can be answered in this way, i.e. by searching for an immediate answer in the knowledge base or by using an inference rule, forward or backwards to deduce a conclusion from a group of known facts.

This has illustrated a simple way in which a knowledge base might operate using a logical representation of facts. A great deal of interest in logical methods has arisen recently due to the introduction of the AI programming language PROLOG. This novel programming language accepts logical statements very much like those we have seen and provides automatic inference mechanisms for producing deduced facts. PROLOG also provides various methods for controlling the processing of the logical statements and so is more flexible and more general in application than theorem provers.

Let us now consider the various features of this type of representation.

1 It is a simple economic notation that is based on mathematical ideas. It is clean and fairly pure in its construction.
2 It has a solid foundation and is well understood with many years of mathematical literature supporting the notation.
3 Inference mechanisms for information retrieval and problem solving are available and these are well understood and are under active research and development.

4 Predicate calculus can be extended to higher logics that can represent space, time, intention and other forms of knowledge.
5 Although logic can be used to model any other knowledge representation technique it is debatable whether this is an efficient or effective approach.
6 There is no inherent organization in the knowledge base. This can be a problem as hierarchies and other types of groupings or associations are not readily available but have to be built in by adding extra structure to the logical system. One way of doing this would be to add extra arguments into the predicates that would define the group or hierarchy to which each predicate belongs. There is some controversy about the best methodology here and different techniques are still being proposed.
7 If a knowledge base is not organised (as suggested in 6), then every statement is available as a candidate for examination by the inference processes. In a large knowledge base this could lead to enormous processing load.
8 It is not easy to formulate procedural or heuristic rules in logic. Logic is essentially passive, consisting of statements which are true or false, and recipes for actions or processes can not be represented in a natural manner.

5.2.2 Networks

Networks are a very straightforward idea. They are often called *associative* or *semantic nets* and consist of two essential parts: *objects* (or *concepts*) and *associations* (or *relations*). These are frequently drawn as a graph with objects drawn as nodes and the associations as links between the nodes. The first three axioms from our previous example, drawn as a net, would look like Figure 5.5.

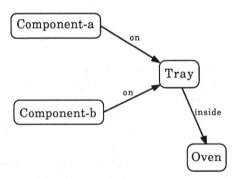

Figure 5.5   Three axioms as a network

Usually there will be much data in an associative net because there will be connections from each node to all other relevant factors and all other associated items. So a network is usually quite a large and complex structure. For example, Figure 5.6 might represent just part of the knowledge base for the heat-treatment plant.

The associations between the nodes can be of different types in order to allow different kinds of relation to be established. Some common associations are labelled in the diagram. Thus 'inst' represents an instance or an element of a class of things, e.g. 'robot' is an instance of the class 'handling-equipment'. 'Is-a' is the link for the generalization of a class,

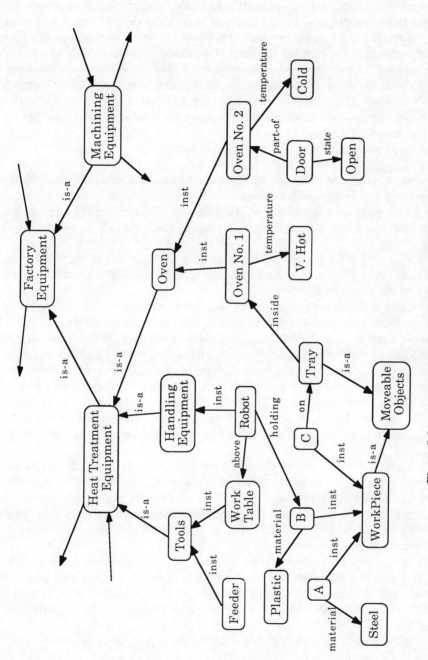

Figure 5.6 Associative network for heat treatment plant

hence: 'heat-treatment-equipment' 'is-a' member of the class 'factory-equipment'. 'Part-of' is an aggregation link which relates various component parts to a parent super-system. Other types of link indicate states, properties, and spatial and physical relationships.

By examining the diagram it is quite easy to learn about the application area simply by looking at nodes and following and reading the relevant links. There can be quite a lot of knowledge embedded in such net structures. Unfortunately, the ease with which networks can be read can lead to over-optimistic assumptions about their 'intelligent' abilities. The *name* of a label can all too easily suggest deeper knowledge that is simply not present in the system's level of understanding. An additional problem is that although there exist mathematics for network structures these have rarely been used in AI and no general notation or usage has developed.

The main features of associative nets are as follows:

1. They are easy to comprehend. They appear to have a natural and straightforward relationship with human understanding.
2. The access paths are easy to follow. By following the links and pointers we discover other facts that relate to a given node without searching through the whole knowledge base.
3. There are grouped associations. All the links from one node give all the relevant data for that node. This gives a natural indexing mechanism and improves on logic systems which do not have associations.
4. Contexts or partitions are possible. If we find that a particular node needs a great deal of data we could have a special link to another small network describing that node. This creates a hierarchy, and it will be very easy to invent our own links to represent branches up and down between hierarchies. Sometimes these are known as contexts and we talk of context switching when we jump from one layer in an associative net to another.
5. Properties can be inherited from higher connected nodes. Note how oven-1 and oven-2 are instances of the class 'oven' (see Figure 5.6). Any general properties of 'oven' can then be inherited by the actual ovens. Similarly, 'heat-treatment-equipment' might be designated as being indoors and so this property could be inherited as well, via the is-a links.
6. We must beware of the danger of reading too much out of our networks. Just because impressive messages can be generated does not mean a network fully represents the semantics of the application area.
7. There is no standard terminology. The mathematics of graphs has largely been ignored in AI and designers tend to invent their own systems. Consequently, most associative nets have significant differences.
8. There is no temporal structure. If we add new facts to the knowledge base then they simply appear in the network and we have no way of knowing which ones were inserted first and how the network developed. The network looks like a snapshot in time. If we wish to facilitate temporal reasoning, we will have to create additional structures and notations.
9. It can be difficult to implement logical relationships in associative nets. For example, to represent the conjunction (AND), of a group of links out of a node would need special additional structures.

Building a knowledge base 81

5.2.3 Procedural knowledge

Now consider our factory oven example from the programmer's point of view. Programmers spend their time designing operations that bring about a desired state of affairs. That is, programming is concerned with operational or procedural information: recipes for action. Of course, procedures and programs are a form of knowledge, but in an imperative form. In other words, procedures contain knowledge about *how to do* things rather than knowledge about *what is true*. So for the heat-treatment plant, we might have a sample robot program as follows:

Place component-*a* on tray
Place component-*b* on tray
Place tray inside oven
Wait 15 minutes
Remove tray from oven

This is a fairly high-level description. The first instruction, 'Place', is likely to be a macro or subroutine name that expands into more detailed code like:

Open gripper
Move to first item location
Close gripper
Move to second item location
Open gripper
Move to rest location

We do not often think of program code as being a source of knowledge because its main purpose is to control a process rather than impart information. However, as an alternative to passive storage, knowledge can be generated or retrieved through a computation, using the conventional execution of a procedure. In addition, if we view procedure source code as a fund of knowledge we see that it contains much valuable data. A clear, well-structured procedure can be read by a person, or intelligent system, to reveal all kinds of application relevant data.

Some features of procedural knowledge can be listed and we notice that several of these highlight differences between declarative and imperative knowledge.

1 Knowledge can be generated by procedures in two ways; either they can be executed so that the results give new information or they may be read for the data contained in the source code.
2 Time is inherently involved. All recipes and instructions contain some sequencing information, either implicitly or explicitly. This gives an ordering to events that can be used in temporal reasoning. Precise time intervals are sometimes included and give even more temporal data.
3 Procedural knowledge is useful where programs of action are involved. Any heuristic guidelines regarding the control of operations are easily incorporated into procedural schemes.
4 Most procedural knowledge is localized. As instructions usually concern only a small part

of the application or environment at any one time, it is not easy to envisage the current state of the system in global terms. The relationship between data within a procedural fragment and global facts outside is not always clear. There are problems here similar to the maintainability issues that arise when controlling distributed databases or maintaining large software projects.
5 It is very difficult to untangle control information from factual data in procedural knowledge. This is is a barrier to analysis and development.
6 If procedural knowledge is to be examined and manipulated by a system, rather than only executed, then the nature of the implementation language is important. The programming language LISP has been heavily influential here. In LISP there are no distinctions between programs and data. LISP can thus either execute or read a piece of data. PROLOG continues with this concept. PROLOG programs can be viewed as declarative logic statements and yet during execution the statements control the flow of processing.
7 In many AI systems the knowledge base is often mixed. Facts often co-exist together with imperative information that describes how to solve sub-goals or use other facts. The implicit use of procedures to generate factual data is called *procedural attachment* or *procedural embedding*.

Procedural knowledge is a very general concept and we will frequently find the idea involved in other topics and integrated into different systems. Perhaps the important point is that our robot task programming languages do contain knowledge about the task and, as well as simply executing the programs, we can use them explicitly so that knowledge will become available for reasoning and descriptive purposes.

We will now look at two major schemes for mixing the declarative and imperative knowledge representation methods. These are called *frames* and *production rules*.

5.2.4 Frames (or schemas)

Frames are groupings of declarative and procedural knowledge that represent distinct concepts or entities. A frame is a collection of related ideas about one particular concept. There can be various frame components depending upon the application, and the actual design of the frame system will be strongly influenced by the nature of the tasks that we are dealing with. Figure 5.7 shows an illustration of a frame with some typical contents. There may be fixed parts in a frame which could hold descriptive data, conditions, relations, or even procedures. Such information can give details on how to use the frame, what to expect during its use, and what to do if something unexpected happens. The variable parts are known as *slots*. These are able to obtain their values by using connections to other frames. Slots usually have names and can have conditions to specify acceptable ranges for their values. Slots can be filled either by obtaining default values, by inheriting values from other frames, or by the action of special procedures known as *demons*. Figure 5.8 shows how part of our heat-treatment application might be implemented in frames. The frame system archetypes are standard skeleton frames which are matched to the conditions prevailing at the time. When the current problem data provide a reasonable match to the structure of one of the archetypes a copy of the archetype frame is taken (called an *instantiated frame*) and the slots are filled with data values. Notice how the different sorts of inheritance are used. The instantiated frame for oven number 2 has three slots. The 'contents' slot inherits values from another frame; in this case, the oven contains the tray and the tray supports the

Building a knowledge base

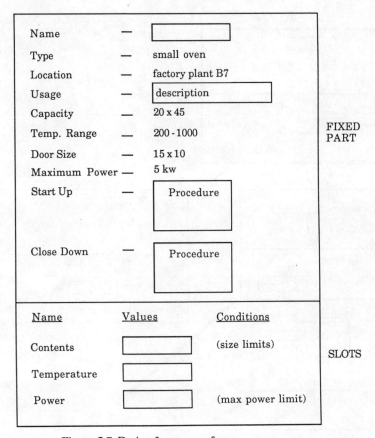

Figure 5.7 Design for an oven frame

components by a second inheritance. But the 'power' slot for the oven is filled by default; as there is no direct frame concerning power the default value is copied from the archetype. Finally, the temperature is produced by a procedure or demon, often known as an *'if needed'* demon. This contains procedural knowledge in the shape of the formula: temperature = 0.2 x power. If the temperature has not been found by other means, the demon can be invoked and the formula activated. Hence, whenever the power is known the temperature can be computed.

Frame systems are based on early ideas in artificial intelligence and psychology. There exists no formal theory and so frame structures and their operation vary considerably with each implementation. The main implementation problems are in matching, that is, in dealing with structural matches between the incoming data and suggested candidate frames. This problem has stimulated various different approaches for handling the common problem of partial incomplete matching. Inheritance can also follow different schemes, e.g. default values might be chosen either before searching for inherited values or afterwards. Consequently, there are many important options open to the designer of a frame system. The main advantage of frame systems is that the procedures and data associated with a particular concept can all be grouped together and made very pertinent for a particular

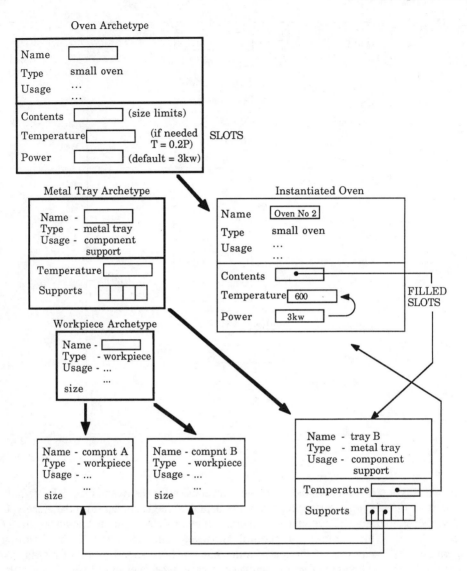

Figure 5.8 Frame instantiation and inheritance

application. Another term for frames is 'schemas', and a very closely related idea is known as 'scripts' (the main difference being that scripts contain some event sequencing data). Frames can be implemented using LISP or PROLOG data structures but the procedures involved in the matching mechanisms are not trivial exercises. Various attempts at frame-based programming languages have been made, e.g. FRL, KRL, KL-ONE, but these have been experimental and none of them has been used very widely. There are now some powerful high-level AI tool-kits available and most of these support the frames idea in one form or another. These tool-kits, such as ART, KEE and KNOWLEDGE-CRAFT, are more than programming languages as they offer comprehensive development environments. They

Building a knowledge base 85

provide a wide range of controls for the various matching and frame-design options mentioned above. Although expensive, such systems may provide a better cost-effective environment for building large frame-based systems than that offered by AI programming languages.

To summarize frame systems:

1. Frames integrate procedural and factual knowledge into modules that contain the essence of an object or concept.
2. Inheritance mechanisms are well developed with several basic methods and a range of options for the designer.
3. Because frames can hold expectations they can support expectation-driven reasoning in a search for confirmation.
4. Frame organization, being based on domain concepts, tends to produce a focus of attention directed towards relevant problem needs.
5. The large module size of frames means that they are not well suited to problems based on large numbers of very small independent rules.
6. Frame-matching algorithms are complicated because they must handle partial and incomplete data and decide which is the best matching frame.
7. Frame systems are not easy to implement, although tools and development environments are now becoming more available.
8. Frames are best for stereotyped situations rather than for novel events. However, variations on a basic theme can be handled and so frames offer potential for representing the ranges of components, machines and processes as are found in industry.

5.2.5 Production systems

Production systems (not to be confused with manufacturing concepts!) are often known as *rule based systems*, and this really explains the idea. A production system is a collection of rules that has been designed to encapsulate the knowledge-processing needs for a given application. All rules are of the form:

IF condition THEN action,

and they are stored together in an area known as the 'production rule memory'. Figure 5.9 shows a schematic diagram of the production system idea. The rule condition parts are matched against data in the 'working memory' where all incoming data arrive. When a match occurs with a particular condition, a rule is said to have fired and the action part of that rule is then performed. Actions are operations on the data in working memory and consist of things like: assert (add to working memory), erase (remove from working memory), replace (change items in working memory), read (from input), write (to output) and stop. The actions in Figure 5.9 are simplified for illustration purposes: in order to perform the action — increase (heat) — assertions would be used to change variables in working memory that are connected to external control signals. All the knowledge in a production system is found entirely in the currently-known facts in working memory and the information contained in the production rules. The idea has its origins in Markov algorithms and grammar replacement systems.

```
                    ┌─────────────────────────┐
                    │ Production Rule Memory  │
┌───────────────────┴─────────────────────────┴──────────────────────┐
│ Rule Number  Condition Part                  Action Part           │
│     1        (temperature low)               increase (heat)        │
│     2        (pressure low) & (temperature high)  increase (flow-rate) │
│     3        (flow rate high) & (temperature high) reduce (heat)   │
│     4        (presure low) & (temperature low)  increase (heat)    │
│     5        (flow rate high) & (temperature low) reduce (flow-rate)│
│     6        (pressure very high)            open (relief-valve) & turn off (power) │
│     7        (temperature very high)         turn off (power)      │
│     8        ...........                     ........              │
│     9        ...........                     ........              │
└────────────────────────────────────────────────────────────────────┘
```

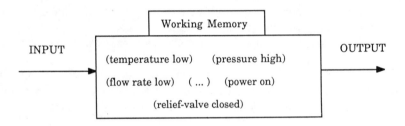

Figure 5.9 Production system architecture

It is worth noting that production systems can be considered as programming languages; designing IF–THEN statements for a given problem is clearly a programming task. But the process of designing frames, procedures or networks is also a form of programming. We will not be concerned with technical distinctions between program structure and organized data, but leave this to the theorists. Our aim is to identify all relevant knowledge and its form of representation or implementation, no matter where it exists or how it may be used.

This form of knowledge base is completely global in the sense that all the conditions are available to be matched against the working memory. This means that many productions may match at the same time and so a conflict resolution method is needed to select one rule to fire from the set of candidates that match. This is known as the 'recognize-act' cycle and is illustrated in Figure 5.10. Three items are shown available in working memory and these match rules 2, 3 and 5. A conflict resolution method might determine that rule 2 should be selected for firing and so that rule is fired and action A_2 is executed. This process is performed by a fairly simple procedure known as an interpreter. The interpreter continuously cycles round the recognize–act cycle reproducing over and over again the match, selection and firing process. This is a totally different method of computer control from conventional programming as it does not have a sequential flow of control. At any time, any of the rules might fire and it is not easy to see beforehand what sequence of events will occur. This kind of system is data-driven; what happens next strongly depends on what

Building a knowledge base

EXAMPLE

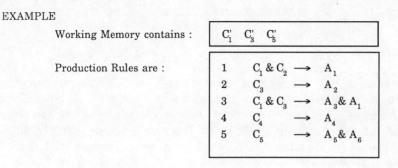

THE RECOGNIZE-ACT CYCLE

(1) MATCHING

    Rules 2, 3 & 5 have matching conditions.
    These form the <u>CONFLICT SET</u>.

(2) CONFLICT RESOLUTION

    One rule is selected from the conflict set :-
    Rule 2

(3) ACT

    The selected rule is executed
    2   ($C_3 \rightarrow A_2$)
    A_2 is performed

The Interpreter continuously cycles until either no rules match or the action 'HALT' is executed.

Figure 5.10   Production system interpreter

information comes from outside into the working memory. However, it is a very appropriate system for a knowledge base as just those rules which are relevant are most likely to fire for the situation in hand. Following our basic example we might have rules such as:

(inside ?x oven) → assert(temperature ?x very-hot)

(on ?x ?y) & (inside ?y ?z) → assert(inside ?x ?z)

With these rules we can produce the deductions shown earlier with logic. These are, in fact, essentially the same inferences we used before. Here we are building conditions using only NOT (~) and AND (&). We use the notation ?x to indicate a special variable that will match any data item. Thus, the data:

(inside tray oven)

will match:

> (inside ?x oven)

if ?x assumes the value 'tray' (we say ?x is bound to 'tray'). At the first occurrence of ?x in a rule, a binding is established, then if ?x occurs again in that rule, it will be replaced by its bound value. Note that ?x and ?y are distinct and do not match the same items. This notation is not standard, and other variations can be found. It is easy to design your own different production-system language and it is also quite easy to write a simple production rule interpreter. More elaborate rules can be constructed:

> (empty tray) & (on ?x ?y) & ~(equal ?y tray) → write('move' ?x 'onto tray') &
>   assert(on ?x tray) & erase(on ?x ?y) & erase(empty tray)

This says 'if the tray is empty and something, ?x, is not on the tray, then move it to the tray by performing suitable actions'. In our system, all actions on the right-hand side of a rule are carried out. Notice that using:

> (empty tray) & ~(on ?x tray) → write('move' ?x 'onto tray') etc. ...

looks more like the English statement, but this rule does not use the object which ?x is standing on, and hence can't update the support relationship very easily. Notice how procedural effects can be introduced here even though the rule base looks fairly passive. In the above example the write statement sends instructions about moving things to the outside world. These instructions would control our robot and cause the tray to be placed inside the oven. Thus, it is possible to program our heat-treatment system through a production architecture and gain the benefits of this style of knowledge base. One advantage is that the rules we are using could be made very explicit and could then be examined by processes *other* than the robot-control system. So production systems have some features of programming and some features of passive database systems. It is worthwhile practising with some examples to make sure you understand the ideas behind this. The best starting point is to trace out fully the operation of a production system. By writing down each step you will see how data-driven systems operate in practice.

The features of production systems are worth listing:

1. It is a very simple design. There is only one rule structure and the basic method is very straightforward.
2. It is a data driven scheme and therefore the execution paths are not easy to follow before run time. It is difficult to be sure that your program will execute quite as expected. This is especially true of large knowledge bases with thousands of rules.
3. Input–output operations are treated no differently from any other actions. This is significant because it means that input-output data can be handled and processed just like an item of knowledge.
4. It has a public control regime. All the rules 'see' all of the working memory and they capture control and then broadcast their results to everyone else. The working memory contains the active goals and intermediate results that are being processed. At any point

Building a knowledge base 89

in time, the working memory is a snap-shot of the 'attention' of the production system.
5 The rules can be read backwards to explain reasoning. If we want to know why an action was performed, we can simply look back through the possible conditions that could have caused it. This is an important feature of a knowledge base, as most intelligent systems are required to be able to explain themselves and justify their assumptions. The ability to re-work the process of reasoning is very valuable.
6 In a similar way as for logic based methods, reasoning processes can use production rules in either forward- or backward-chaining techniques.
7 Production systems are extensible. As rules are independent, growth is possible merely by adding extra rules (they can also be deleted or updated). However, it is much easier to add rules than to know the consequences of the additions and how they will affect the existing system.
8 The production-rule architecture provides a useful experimental framework for automatic programming, learning systems, and debugging and verification systems. Because meta-rules can be built to describe other rules, systems can be designed to reason about their own performance, and, hopefully, improve themselves.
9 There are analogies between production systems and human models of cognition. These have stimulated cognitive scientists to develop a range of advanced production system languages.

Although there are many more details that could be discussed, for example, different methods of conflict resolution and various modes and styles of operation, we must now refer the interested reader to the relevant literature in the 'Further reading material' section on page 97.

5.3 Representation review

We can now review the various knowledge representation methods in terms of their different advantages, shortcomings and capabilities. Remember that the reason for building a knowledge base is to support reasoning processes. We soon discover that such reasoning for our industrial automation domain will involve various types of knowledge. These will include: 'bare' facts and statements; time reasoning, involving sequential events; spatial reasoning, involving routes and navigational problems; and geometric reasoning, involving the relationships between shapes — especially for computer-aided design and automatic assembly. Some of the physical processes will also need to be explicitly understood in order to modify and change parts of the task. Some general physics might be involved, as systems may need to know about gravity, electricity and natural phenomena which might interfere with, or alter, a process. Also, machines themselves will need to be described and reasoned about.

This is a formidable list, but notice, however, that many things are excluded. These are the topics that are normally associated with human reasoning such as emotional states, different methods of understanding language, problems of perception, and cognitive processes in humans generally. Fortunately, we do not have to worry about these too much in the industrial domain although, of course, some branches of AI do deal with these issues.

The basic representation schemes described so far do not offer any special facilities for

	Logic	Networks	Procedures	Frames	Productions
Association facilities	A	✓	X	✓	A
Ease of adding temporal data	X	X	A	X	A
Ease of adding procedural data	X	X	✓	✓	A
Problem structure expressiveness	X	✓	X	✓	X
Extensibility of knowledge base	✓	✓	X	A	✓
Control/knowledge Separation	✓	✓	X	X	X
Rigorous notation and foundations	✓	X	X	X	~
Simplicity of concepts	~	✓	~	✓	✓
Ease of implementation	~	✓	~	~	✓
Debugging and execution comprehension	~	✓	~	✓	X

KEY ✓ good / facilities supported
   X bad / not a natural facility
   A can be incorporated by additional structures
   ~ fair / low

Figure 5.11   Comparison of representation schemes

the engineering domain, rather they are general methods that have to be adapted and extended for particular applications. Consequently, knowledge-based systems for areas like geometric reasoning, for example, will be enhanced with models and techniques drawn from the extensive literature in computational geometry. In later chapters we examine some of these developments, but here we consider more general points of comparison.

In Figure 5.11 we take ten features and list them against each of our five representation methods to produce a table. The features are:

1 The availability of mechanisms to allow the association or clustering of facts into concepts.

2 The ability to record and use time values in the knowledge base.
3 The ability to incorporate procedural data for controlling machines and instructing computers.
4 The ease with which the notation expresses the basic nature of an application problem.
5 The facility to add new axioms to the knowledge base with consistency being maintained.
6 The degree of clear separation between application knowledge and control data.
7 A sound foundation for the representation method and any notation system.
8 The straightforward nature of the design.
9 The ease of implementing the scheme.
10 The clarity of the notation to give easy understanding of operations during their execution and debugging.

Figure 5.11 shows the various strengths and weaknesses of the different systems. We have concentrated on implementation factors, rather than on theoretical distinctions between methods, because these are perhaps more important for the application engineer. It is quite clear that some methods are much better than others for certain types of task, but it is also true that all methods can claim to handle most tasks from a theoretical point of view. Some AI experts argue that, because extensions to one method will be sufficient to cover the features of all the other methods, we should concentrate on developing only one method. Logic has a powerful claim here, because of its ability to model formally the other methods. However, this issue must await further developments in AI.

The choice facing the implementor is very similar to that experienced when choosing a programming language — it is really a question of appropriateness; clarity, ease of implementation and a good match for the given problem matters more than the abstract virtues of a particular method. We hope that the different methods outlined here will be seen as *ideas* that stimulate knowledge base designers and users, and help to clarify the concepts involved.

5.4 Knowledge integrity properties

Now that we have seen five different techniques for structuring a knowledge base, let us next review the various characteristics of knowledge itself. As most of these properties, in some way, concern the quality of factual data, they tend to introduce measures that are used to indicate the level of integrity of the data.

5.4.1 Incompleteness

A knowledge base is nearly always incomplete — some pieces of vital, significant or desirable data are always missing. Notice that complete knowledge would usually allow us to develop an algorithmic solution to most problems, as we could compute an answer from a set of relevant information, given enough time. Incompleteness is thus a major characteristic of knowledge-based systems and distinguishes their heuristic approach from algorithms and database retrieval techniques. Incompleteness has serious implications for the design and organization of a knowledge base (and also for the knowledge acquisition process). If we give a knowledge base the query:

(made-of bolt-5a brass) ?

and if there is no stored data on bolt-5a or brass in the system then we will simply get a nil answer. This nil response does not imply negative information; it does *not* mean that bolt-5a is not made of brass. It simply means that the system does not know the answer. This is different from conventional databases, which usually operate on the basis of complete information. In this case, known as the *closed-world assumption*, everything that is true for the given application world, is stored in the database. Thus, any query that does not generate a true response can be assumed to be false. In other words, if:

(made-of bolt-5a brass) ?

gives the answer nil, and if the world is closed for that particular database, then bolt-5a is *not* made of brass. However, whenever the data are incomplete this convenient assumption breaks down and we have to adopt heuristic tricks to solve the ambiguity. One way of handling incomplete data is to add an extra truth value into the system. Thus, we could have a value, say 'unknown', so that any query could now respond with true, false or unknown. Knowledge in the system could also be tagged with any of these three values. Of course, the whole knowledge base will now require special procedures to be designed and built to handle this feature. Such schemes account for a great deal of research effort in AI.

5.4.2 Inconsistency

This refers to the situation where two logical statements are in either direct or indirect conflict. For example, we might have the rule:

(colour ?x red) & (made-of ?x metal) → (temperature ?x hot)

and the statements:

(colour tray red)

~(temperature tray hot)

If we later find out that the tray is made of metal then we can logically infer either the tray is hot (using the inference rule) *or* the tray is not hot (using the given axiom) and we have an inconsistent system. Because the system of logic breaks down at this point, it becomes possible to generate all sorts of invalid conclusions. It is a property of logic that *any* desired conclusion can be proven with an inconsistent system. This is another major problem because, unlike databases, knowledge bases might well contain much inconsistent data drawn from complex and ill-structured application areas. One solution is to attach a special indicator, e.g. the label 'inconsistent', to every fact and rule that is known to be in conflict. This separates the consistent parts which can still be used in logical deductions. A further development is to expand the indicators into a range of 'plausibility' values (say real numbers between 0 and 1.0) that define the 'degree of inconsistency'. Special machinery has to be created in order to reconcile inconsistent facts in a large system.

5.4.3 Inaccurate data

This occurs when the data in the knowledge base do not match the real world. For example we could have a fact asserted:

    (oven-1 switched-on)

when this is not the case for the real oven. It can be difficult to detect such inaccurate facts from inside the knowledge base because they are not necessarily inconsistent with the rest of the knowledge. We can only rely on the input of valid data via sensing systems and other inputs to track down inaccurate data. Of course, sensors themselves are a major source of inaccurate data but methods exist to reduce noise and increase reliability through corroboration techniques. A good strategy for identifying and correcting inaccurate data is to link up to the processes trying to reconcile inconsistencies. When an inconsistency is to be investigated, the first action should be to use all relevant data to drive a search for inaccuracies. Requests for new data or sensor readings might then bring new information and this may well solve both problems of inaccurate and inconsistent data.

5.4.4 Uncertainty

Uncertainty is the degree of truth of a particular statement. There are various different methods; for example, we can arrange the certainty of a truth value to be measured in the range 0 to 1 and a confidence value can be assigned, also with the same range, 0 to 1. By these means we can attach values to our statements indicating their likelihood and our confidence in the truth estimate. For example, if the error rate for a particular machine is estimated to be one product in twenty, we could give a statement representing this as;

    (product item-x faulty (0.05, 0.0))

This indicates our estimate of the probability of a particular instance of item-x being faulty but shows that we do not have any confidence in it. After a large run on the machine we may find by experience that the error rate is, in actual fact, 1 in 20 and then we could assert;

    (product item-x faulty (0.05, 0.99))

Both the certainty and confidence values can be changed by logical reasoning processes. Uncertainty reasoning is the method used in many expert systems for weighting evidence and combining statements to produce more reliable data and conclusions.

5.4.5 Imprecision

Imprecision is not the same as uncertainty. It concerns vague statements or inexact measurements. For example, we could state;

    (temperature oven high)

But the question is — how high? This is a vague statement, compared with

>(temperature oven 296.5°C)

One method of expressing degrees of imprecision is to use *fuzzy sets* in which membership of a property set is given by a fuzzy variable, usually ranging over 0 to 1. Thus;

>(temperature oven high (0.7))

implies that the membership of temperature is 70% in the high range. Probability distributions have also been used to represent imprecision. There is a large literature on fuzzy logic and there are many methods for combining such values to give more reliable data from fairly imprecise raw facts. These methods have been used to improve the performance of sensors and for combining and merging information from different sources. There are confusions and controversies in the literature about the use of fuzzy logics, and uncertainty measures and probability logics are often confused with fuzzy logics. The best use of fuzzy logic seems to be in representing imprecision in measured quantities.

5.4.6 Non-monotonic reasoning

In most logical systems if propositions A, B and C are true then adding another true one, D, will enable combinations with D, like (A & D) and (B v D), to be true also. In other words, the system is *monotonic* in its growth because adding an extra piece of information increases the total amount known. But in a *non-monotonic* system adding D could actually lead to less knowledge because it might cause, say, B to become invalid and be removed. An example of this sort of problem is seen in default reasoning. Let's imagine we have a component database which assumes that if a component has no data entry, it is made of mild steel. So, unless stated otherwise, all objects are made of mild steel. Now, if a query asks for data on a particular bolt and there is no prior entry for it, a default fact will be generated;

>(made-of bolt-x mild-steel)

Later on, we may find that this bolt is actually made of brass, in which case we would have to withdraw the first statement and insert;

>(made-of bolt-x brass)

Notice that we would also have to withdraw *all* statements created from deductions based on the original data. In such non-monotonic systems we will need special methods to record all these dependencies and to keep track of potential conflicts and maintain consistency. The recently-developed techniques of truth maintenance are now being used in *truth maintenance systems* (TMS) so that knowledge bases can provide just this kind of support for complex dynamic reasoning processes.

Building a knowledge base 95

5.5 Organization and control

There are many design issues that do not concern the quality and representation of knowledge but refer to the general operation of the knowledge base. As in most of AI, these topics do not exist in isolation but impinge on many other aspects of the structure and design of a system. The aim of this section is to introduce just a few of the many issues.

5.5.1 Access and matching

Knowledge bases, like databases, require good access mechanisms in order to retrieve factual data. Both need easy methods of access and association but knowledge bases have the additional problem that retrieval must also work for partially known facts. This means that access involves a matching process; we might want to find a given fact in a knowledge base from a partial pattern or specification, or we may want to retrieve all data, however fragmented and incomplete, that relates to a given query. Matching is a powerful technique that can be used in various ways: to classify things into types, to confirm that something exists in a knowledge base, to decompose goals into sub-goals, to reduce input into smaller portions and to correct errors in patterns. Matching also occurs at different levels — at the *syntactic* level (where we are only concerned with the form of the data); at the *parametric* level (where we focus on the variables involved); and at the *semantic* level (where we are interested in the function of the item that is being matched). There exist matching algorithms for all the main representation methods described above and these are discussed at length in text books.

5.5.2 Multiple representations

Data are often used in various different ways for different purposes. The information handled by an engineer will be of a different form from that used by a manager although both may be working on the same product. It is sometimes sensible to maintain two or more representation systems side by side in order to cater for different styles of access and inferencing. This might seem less efficient than the alternative of a unified system with suitable conversion schemes, but we shall see that there are considerable benefits to be gained by multiple systems. The interaction between multiple representations can create the framework for powerful additional reasoning features. Multi-layered knowledge bases are now very much under active study and may well be the key to solving some of the difficult problems which arise with a single unified representation scheme.

5.5.3 Self-knowledge and meta-knowledge

There are many kinds of self knowledge, often called *meta-knowledge* or *meta-rules*. The basic idea is to create and maintain a set of special facts that describe the features, structure or operation of other facts. An immediate example can be seen in the type definitions used in modern programming languages where the types of different variables are explicitly defined. Other kinds of self-knowledge are necessary for producing explanations. Questions like 'why did you do that?', or 'how did you do that?', require some form of self-knowledge in order for a system to be able to give sensible answers. Meta-knowledge can also be used to control processing; for example, in production systems a group of meta-rules might give guidance when selecting rules from the conflict set. Research is active in many areas of meta-knowledge representation. It is a fairly easy matter to implement meta-rules in declarative schemes, by introducing a new class of rules or predicates, but it is not so easy to do this with more conventional programming languages. This is because most programs are intended only for execution, not for analysis and inspection by other programs.

5.5.4 Belief systems

One of the latest knowledge base developments is the idea of *belief systems*. Such systems incorporate models or mechanisms that distinguish what is *known* to be true from what is *believed* to be true. Something which is known is a 'fact'; it is known to be true because, if necessary, we can justify it by supplying other facts or arguments as support. But other items of knowledge are also treated as true facts even though they are not justified; these are 'beliefs' as they are simply believed to be true. There is a third type of fact called a 'hypothesis', which is not yet known or believed to be true but has some justification for why it might be true. Thus, there are three possibilities: true facts with justifications, unjustified beliefs held to be true, and unproven hypotheses with supporting justifications. Special kinds of knowledge-base design are being developed to incorporate these ideas of belief and hypothesis. These *belief maintenance systems* (closely related to TMS ideas) are able to record justifications for each piece of knowledge and modify the status of facts as new information is received. For example, a new fact, A, may add support to the justification of B, thus changing B from a belief to a known fact, while A might also weaken the justifications of hypothesis C and belief D. Justifications are created by linkage mechanisms that specify how a given fact provides support for another fact.

5.5.5 Relative knowledge

Any one particular agent, robot or program will collect a different set of facts about its environment from those known by another agent, even though they may both exist in the same world. Early AI work concentrated on single cognitive agents and so the system's view of the world was the same as the global view held by the designer. It has taken some time to appreciate that there is no global view; a statement about the world will vary depending upon who you ask. One agent will say one thing about a topic while another agent will say something else. This situation, of a set of subjective agents without an objective global view, requires additional structures in a knowledge base. We now have to distinguish our own knowledge from knowledge about what we think other agents know. Such knowledge is important during conversations and for reasoning about the intentions and actions of other agents in the world.

5.6 Summary

As summaries of the individual methods have been given in the various sections, we list here some of the more general observations.

1. Knowledge bases store information in declarative or imperative forms, or, more often, as a mixture of the two.

2 Aspects of knowledge organization include both the relations between facts (structural) and the properties of those facts themselves (integrity).
3 Most representation schemes have strong affinities with other methods. There are some similarities and common threads between the two main approaches: logic → nets → frames, and axioms → rules → procedures. In theory, logic can be used to implement most of the other systems.
4 Schemes of representation range from the formal and rigorous to the informal and innovative. There is no consensus as to which is best. The choice of representation is best made by carefully selecting the desired characteristics, and attempting to match these to the needs of the application.
5 There are many ways to view the problems of knowledge and its organization. The classifications presented in this chapter are offered as aids for analysis. Other authors will present quite different views, e.g. by considering all representational issues in terms of programming language requirements.
6 Current knowledge acquisition and elicitation methods are not yet well developed nor appropriate for many engineering applications. However, sources of knowledge can be located for most engineering processes. These include robot task programs, component and machine design data, sensor configuration data and process physics information.

5.7 Further reading material

Basic material on the representation methods described here can be found in many AI textbooks; the general references cited in Chapter 1 are recommended.

A valuable collection of classic papers is *Readings in Knowledge Representation*, edited by R. J. Brachman and H. J. Levesque (Morgan Kaufmann, 1985). A special issue of the journal *IEEE Computer* (October 1983), volume **16**, number 10, was revised and reprinted as a book: *The Knowledge Frontier: Essays in the Representation of Knowledge*, edited by Nick Cercone and Gordon McCalla (Springer-Verlag, 1987).

A review of the nature of knowledge and the important features that concern AI is: 'Knowledge Representation: Features of Knowledge', by James Delgrande and John Mylopoulos in *Lecture Notes in Computer Science*, number 232, edited by W. Bibel and Ph. Jorrand (Springer-Verlag, 1986).

In an attempt to improve representation schemes to handle considerations like necessity, possibility, belief, time, and relative knowledge, AI workers have explored the potential of some of the more exotic non-standard systems of logic. *Logics for Artificial Intelligence* by R. Turner (Ellis Horwood Ltd., 1984), is a good review text giving details of modal logics, temporal logics and fuzzy logic.

The original frames idea is described in a readable article by Marvin Minsky entitled 'A Framework for Representing Knowledge', in *The Psychology of Computer Vision*, edited by P. H. Winston (McGraw-Hill, 1975).

A collection of papers on production system topics is *Pattern-Directed Inference Systems*, edited by D. A. Waterman and F. Hayes-Roth (Academic Press, 1978).

Readers interested in AI programming languages should consult Ivan Bratko's excellent book *Prolog Programming for Artificial Intelligence* (Addison-Wesley, 1986). A good starting point for LISP programmers is *Common Lispcraft* by R. Wilensky (W.W. Norton, 1986).

Chapter 6

Machinery for thinking about actions

In each action we must look beyond the action at our past, present and future state, and at others whom it affects, and see the relations of all those things. And then we shall be very cautious.

Blaise Pascal

6.1 Introduction

In Chapter 4 we looked at the problems of perception, that is, the cognitive problems associated with sensory processing. Perception can usefully be sub-titled 'thinking about sensing'. In this chapter, we will look at the problems associated with 'thinking about actions'. This kind of thinking is also known as planning as it usually involves the creation of a plan of action. A plan is simply a series of operations or instructions that, when executed, will change a given starting situation into a desired or goal situation.

As a case study, we will concentrate on a model of a robot system designed to move objects in a simplified world. This will form the framework for our discussions of planning systems. We view planning as a broad category of techniques which have the common aim of producing programs for the control of robots and other machines. Current programming methods for robotics involve the writing of a program in which the code explicitly states *how* to do the job, e.g. 'move to position A, grasp object B, move to position C', and so on. In future programming systems, *what* is required will be specified, rather than *how* it is to be achieved. In such cases, we might say, 'put the car battery inside the car', or, 'assemble the water pump'. These very high level instructions would be processed by a cognitive system that effectively generates a *plan* to achieve the desired result. The plan would then be translated into a sequence of steps, thus forming a robot program.

Various methods of creating a plan are possible. All of these methods have their own advantages and shortcomings and we must be very careful in selecting the right techniques for the problem at hand. In particular, we must identify the exact nature of the 'intelligence' problem so that we don't apply an AI overkill in trying to solve much more difficult problems than we really have. In this chapter, we will use a very simple model of a robot

Machinery for thinking about actions

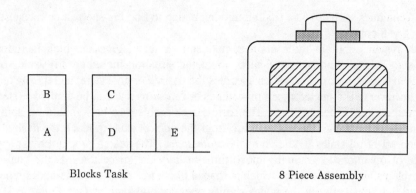

Figure 6.1 Idealized and real worlds

system which consists of a movable hand, and a series of identical components called 'blocks'. The goal of the robot system will be to pile up blocks into simple configurations. Figure 6.1 shows this idealized model contrasted with a more realistic example of an industrial assembly. This *blocks world model* has been widely used in AI systems to demonstrate the ideas involved. There is a strongly implied assumption in much AI work that such blocks models are a good approximation to real world events and are sufficient to support theories that can later be applied to assembly and other mechanical handling problem areas. However, we believe these assumptions are not valid for real-life robotics and, consequently, the model has severe limitations. This difficulty will be discussed in later chapters.

The intention of this chapter is to introduce some basic methods that are fundamental in AI as techniques for reasoning and thinking about actions. We will look at several main categories. First of all, we look at the technique of searching through arrangements of alternative options. This is a major area in AI with a good grounding in research, both in theory and practice, and we will illustrate examples including the application of search in a blocks world planning problem. Secondly, we will examine a technique known as 'goal-directed programming'. This approach involves collections of procedures that attempt to break up a problem and then solve the sub-problems according to their own functional capabilities. Thirdly, we will look at rule-based systems and see how production rules may be used to generate plans. Lastly, we will examine a more generalized idea called 'blackboard systems'.

6.2 Searching for solutions

The idea of mechanically searching through a series of possibilities is a very old one, older in fact than computing, and this concept has a long history of research and development. Two well-developed fields are logic-based theorem-proving and game-playing (where much effort has been spent in recent years on developing chess-playing programs for world tournaments — encouraged by substantial prizes!). Other areas include the solution of geometric configuration problems, many forms of symbolic and other puzzles, and path-finding techniques such as route finding through a map to find the shortest or cheapest path between given points.

Search systems contain three main components: a set of *givens*, which includes the current state and all known useful facts about the situation including any 'rules of the game'; then a *goal description*, which specifies the desired or goal state; and thirdly there is a set of *operators*. The operators are procedures or functions that can be applied to states to transform them into different states. The basic search problem is: how should we apply the operators so as to change the system state from the current state to reach the desired goal state? The set of all states is known as the *state space*. The search program must apply a sequence of operators to gradually move through the state space towards the goal state. Most problems have a state space which is shaped like a tree or graph. The nodes represent states and the links between the nodes signify operator applications (see Figure 6.2). The two most common forms of state space are those known as OR graphs, used for the solution of path problems as in figure 6.2, and AND/OR graphs, which are used for solving sub-goal problems. Figure 6.3 shows an example of an AND/OR graph. Notice how the nodes here represent sub-goals that must be satisfied to achieve higher level goals. This type of search is often called *goal reduction*. The differences between these two search spaces are due to

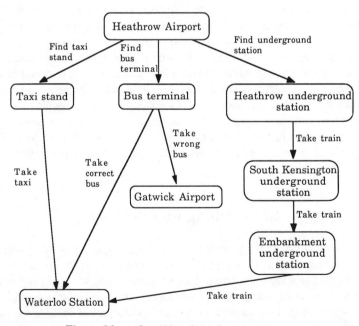

Figure 6.2   Searching through a state space

Machinery for thinking about actions

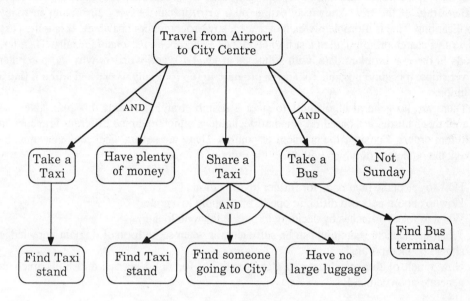

Figure 6.3 An AND/OR state space

the type of relationship that occurs at the node. With a group of OR branches *any* branch can satisfy a node (e.g. find any path to a goal), while for AND branches *all* of the branches must be satisfied (e.g. all of the sub-goals must be achieved).

There are different shapes of search path. A deep path involves many applications of operators, and is usually the result of a search for a specific goal *node*. A shallow path search, on the other hand, often involves the evaluation of many different routes to reach one or more goals. In this case, the best *path* is often the solution or the result of interest, not the goal node itself. The output from a search routine can be either just a statement that the goal exists, or it can be the details of how to follow a path to the goal. For path searching, either the first path found will suffice, i.e. any path, or it might be required that the best path is to be found.

The usual approach to search is to start with a given state, then apply operators to generate another level in the tree, and then repeat the whole process recursively   the next level. However, operators can be applied in different ways. In forward reasoning, the search moves continuously forward from the start state until it reaches the goal, but in backward reasoning, the goal state is examined first and the search works back towards the start state; this is sometimes called *goal simplification*. In a third method, called *means-ends analysis*, operators are applied according to their priority in reducing the difference between the start state and the goal or any sub-goals which may be known in the tree.

In all of these methods there is a nasty problem known as the *combinatorial explosion*. This is the very serious and real numerical explosion which occurs when the branches of any non-trivial tree are enumerated. Consider that there are B branches from each node in a tree, then to find a goal at a given level, say level L, the number of nodes we need to examine is B^L. If the number of branches from each node was 30 and the number of levels was only 5, we would find that the complete tree involves over 24 million nodes! The rapid growth rate of the tree, known as exponential growth, can be very surprising in many applications. This is a problem which catches many programmers unawares. It is quite easy to write a search program, then test it to level 2 or 3 and obtain reasonable results. Then one feeds in the real problem data, with solutions at level 10, and wonders why the computer never stops. It is quite possible for such a program to run for many weeks and still not find a solution.

There are no general algorithms to steer a search straight towards the goal states, so search mechanisms are based on standard techniques which examine different branches in different orders. Consider the problem of control. There are several decisions that must be taken, the outcomes of which will effect the order of events:

1 How to select the next node for further investigation.
2 How to choose between different operators to apply to a node.
3 When to reject branches by deciding to pursue them no longer.
4 When mechanical searching can be sufficient, or when some form of domain knowledge should be incorporated.
5 How much of the search tree to store explicitly for rapid access, and how much to generate upon demand.

These ideas are illustrated in Figure 6.4. Starting at node A, we might have discovered its children B, C, D and E, and wish to proceed further. Choosing between the nodes corresponds to decision number 1; let's select C. Then selecting one of the operators, p, q, r, or s, will lead on to different new nodes: decision number 2. Perhaps operator r might be selected, thus leading to the generation of node F. If, however, any node was considered unlikely to lead anywhere useful, we could reject it and so remove a whole branch of the

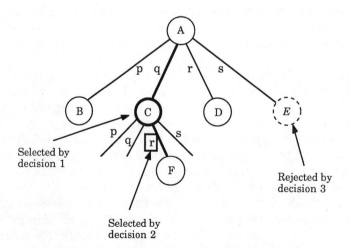

Figure 6.4   Choice points in search control

tree in one stroke, e.g. node E. This decision (number 3) is a risky one as there might be unexpectedly valuable nodes deeper down. Decision 4 refers to the use of application-specific knowledge. For example, if we know that operators p and q are not effective below a certain depth in a given application then we can save search effort by ignoring them. Finally, the last decision is not really a search consideration but concerns implementation trade-offs (although the order of storage or access can sometimes influence the search result). All the variations between the different standard search algorithms arise out of different choices for the above five fundamental decisions. We will start by looking at blind search techniques, that is, those that do not use any domain knowledge and are based on purely mechanical search procedures.

6.2.1 Hill climbing

Hill climbing simply involves choosing the best successor of the current node being examined. Starting with any given node, all of the successor nodes are generated, then they are evaluated by some means to decide which one is the best, the best one is chosen, and the whole process is repeated on the successors of the best node. This technique requires some measurement or evaluation method to decide which successors are best. The nature of the evaluations will be completely determined by the nature of the application and must be programmed for each different searching problem. The evaluation is likely to be heuristic in nature as a thorough analysis of the nodes may be too difficult or too expensive to compute. Hill climbing is a local method, in the sense that it does not take account of global information but simply moves 'inch by inch' towards what it thinks is a better solution, hence the title 'hill climbing'. Unfortunately, it is irrevocable because it always moves forward and has no means to retrace its steps from poorly valued nodes. This means it will wander around aimlessly on flat plateaux (where all the values are similar), or it can get stuck on a local maximum and not find any higher existing maxima. However, the method demands very little in the way of memory and computational overheads, as it simply remembers the current successors and the current path it is working on. Also, it is a non-exhaustive technique; it does not examine all the tree, so its performance will be reasonably fast. This is one of the most primitive search methods, and should really only be used where the search domain is known to be monotonic, that is, where there is a single global maximum. As this is rare in practice, we try to improve our search control by some form of recovery ability. This means making tentative decisions which can later be revoked. Another technique is to use a wider scope of evaluation at decision time. In other words, instead of looking at only one node and choosing its successors, we could look at a range of nodes, and choose the best from those. Most of the following techniques involve these two improvements.

6.2.2 Depth first

Depth first search involves following a path from one successor to the next before looking at its neighbours. Figure 6.5 shows a depth first search in progress. The procedure generates a successor of the given node and then immediately pursues the successor of this. This continues down one branch, until either some processing limit or the goal is reached. The limit is usually a maximum depth rule. When the depth limit is reached, the system

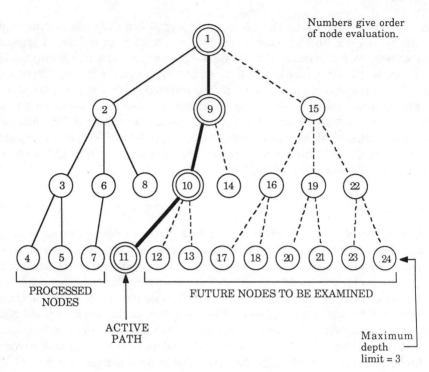

Figure 6.5 Depth first searching

backtracks up to the last successor, and chooses an alternative. In this way, it sweeps across the search space, eventually covering all the nodes within the particular depth limit. The active nodes under consideration form a single path from the root (see Figure 6.5). This search pattern allows nodes to be discarded after they have been fully examined and future parts of the tree do not have to be created until needed. This means that there are low memory overheads because only a small part of the tree has to be held at any one time, and, in addition, the tree can be generated in the same order as it is explored. For these reasons, depth first search is easy to implement and has become a popular computing technique. It is one of the fundamental methods in computer science.

Depth first is a full state space search in that every node is processed, i.e. it is exhaustive within its set limits, a major improvement over hill climbing. However, it has the serious problem that should a branch of a tree be infinite, in the sense that it is possible to apply operators to its nodes *ad infinitum*, then, if there isn't a maximum depth limit, this search technique will spend all its time exploring one particular branch and not coming back to deal with the others. On the other hand, the use of maximum depth controls is also a limitation because the solution that is required may be just beyond the maximum depth. This is called the *horizon effect*, and states which are beyond the maximum depth are considered to be 'over the horizon' and effectively don't exist during that particular search. Despite these caveats, depth first search is widely used in many areas of AI and computing and we will encounter applications in robotics.

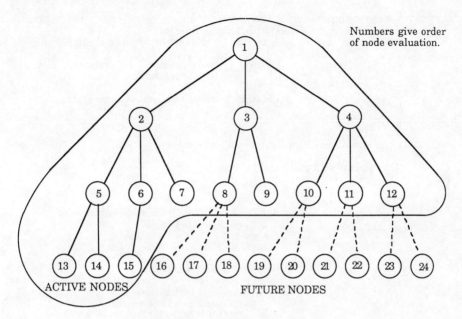

Figure 6.6   Breadth first searching

6.2.3 Breadth first

Figure 6.6 shows a breadth first search in progress, where all the successors are examined at each level before any nodes on the next level are processed. This means that the first goal to be found will be the one nearest to the top of the search tree. Consequently, the problems with infinite trees do not affect breadth first. However, it has very high memory overheads, as it must save each level of the tree in order to generate the next level. This could be a serious problem; often the enormous amounts of storage required rule out this method as a practical technique. Like depth first, this method is also a completely exhaustive search as it will process every node. Because of this, and the pattern of level-by-level searching, it will guarantee to find the shortest path solutions. If there are several paths to a goal and one is shorter than another, then the breadth first method will find the shortest path first. This is one method that *guarantees* some performance behaviour, i.e. it will find a solution, if one exists.

6.2.4 Beam search

This technique is a modification of breadth first searching. At each level, a small number of nodes are allowed to be expanded and pursued to deeper levels. For example, we might choose the best five at each level, and just pursue five down each time. This reduces the amount of search involved, because a large proportion of the tree is now ignored. However, as it is non-exhaustive in this way, it is a *hazardous* process because a goal state might be missed. The quality of the results will depend upon how well we can select the small set of nodes for examination at each level. Such selection decisions will involve some form of heuristics or domain information.

Node Evaluation Function **F = G + H**
i.e. (node cost estimate) = (cost of path from root)
       + (distance-from-goal estimate)

Two lists of nodes -   OPEN   Nodes generated and evaluated but not expanded.

                       CLOSED   Nodes with successors generated

ALGORITHM

1) Put root in OPEN together with its F value (use F = H)

2) Repeat until goal found
       if OPEN empty
           exit (FAIL)
       else
           select BEST node from OPEN (i.e. node with lowest F)
               and move it to CLOSED
           if BEST = goal
               exit (SUCCESS)
           else
               generate successors of BEST
               for each successor
                   set pointer back to parent
                   set G (successor) = G (parent) + g
                   if in OPEN or CLOSED
                       ignore node (repeat - already seen)
                   else
                       compute its F value and put in OPEN

Figure 6.7   The A* best first search method

6.2.5 Best first algorithms

There are various techniques known as best first searches. These are used in cases where there is a means of calculating the cost of different paths and methods for estimating the merit of various choices in the state space. Consequently, this type of search can, in some sense, find an optimum path. The *A* algorithm* is one of the best known methods for searching OR graphs and is shown in Figure 6.7. The algorithm maintains two lists of nodes: the open list, which contains generated nodes that have been evaluated but not had their successors expanded, and the closed list which contains the completely processed nodes, that is, the ones where the successors have been generated. An evaluation function is used to produce an estimate of the cost of any node in terms of its distance from the root (a precisely known quantity) and its estimated distance from the goal (a heuristic measure). These two quantities, G and H, are added together to produce a combined cost function F. Nodes with the lowest values of F will be nearer the goal and have shorter paths from the root. Thus, F summarizes how well the search is doing. It is important that H should always tend towards an underestimate of the real distance from the goal as this can be shown to lead to an optimal solution.

In operation, best first searches act like a kind of mixed depth and breadth first search. They look for the most promising branches of the tree, and pursue those by expanding their successors and selecting the best ones for the next round. Notice that each node is checked for duplication against *all* others in the open and closed lists. This check means that any shape of graph or tree can be effectively searched as repetitions of nodes and sub-trees will

Machinery for thinking about actions

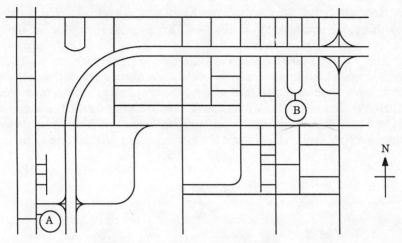

Figure 6.8   A road map for route planning

be detected. Only methods that store lists of all encountered nodes will be able to detect repeated nodes or cycles in a graph. There are interesting special cases for the evaluation function, for example, if g (the path cost increment) is set to zero, then 'any path' solutions are found, as route differences are ignored. If g is set to 1, as is usual, then paths with the fewest steps are found. If g is computed from some given cost function, then the method will find the cheapest path to the goal. If the estimate H is set to 0 and g is set to 1, then the search reduces to the breadth first technique.

There are many variations of best first search designed for all kinds of search space. The facility to tune the parameters offers extra scope and there exists much practical experience in using the method. The method is still dependent on domain heuristics for the cost estimates, however, and so its hazardous characteristics must not be forgotten.

Let's now look at some examples of search and see how this technique can be used to make plans.

Example 1 — A navigation problem
Figure 6.8 shows a road map and the problem here is to find the shortest map route from point A to point B. But notice that 'shortest' can have several meanings. It could mean the shortest in geometric distance, or it could mean using the least complicated directions from A to B, or using the fewest number of junctions, or using the fastest trunk roads. These are all domain-dependent aspects which must be incorporated into the problem specification if the search is to perform effectively. In our case, let's assume that the shortest geometric distance will suffice. If we now turn to a standard text book and look for advice on how to solve this problem, we will most likely find that breadth first is recommended. The argument given will be that we should minimize the worst case possibilities. A worst case solution would be to branch off in the wrong direction and explore many streets, eventually going to another town or even another country before returning to go off in a better direction. This could easily happen with depth first search which appears to offer the worst possible behaviour that might occur! Notice also the structure of the state space in this

problem; there will be a relatively small number of nodes (road junctions) in the solution, but many different solution paths are possible. This suggests again that the shortest path behaviour of breadth first is appropriate.

However, if we use a little domain knowledge, we see that this is not really a difficult problem. In domain knowledge terms, we would use a rough sense of direction. We know that B is 'roughly north-east' of A, and so if we found ourselves going west for any length of time, clearly this would be wrong. Such simple knowledge, i.e. a crude sense of direction, can have dramatic effects on performance. Some results are shown below for an implementation of this particular problem using both breadth first and depth first searches.

Search	Number of nodes processed in map problem	
	single ended	*double ended*
BREADTH FIRST	904	784
DEPTH FIRST	509	450

In these experiments a directional bias towards the goal was added to the depth first searches. In other words, a sense of direction was added so that going in the approved general direction was of higher value than going in other directions. The search process was performed for both single-ended search, branching out from A, and for double-ended search, branching out simultaneously from both A and B and meeting in the middle. The results are interesting in that depth first search was actually much faster than breadth first. This is because breadth first fans out evenly and considers a large number of local nodes, whereas depth first explores deeper node sequences before expanding sideways. Figure 6.9 shows how this effect works. Clearly a sense of direction is a very helpful heuristic for guiding this kind of search.

There are other techniques which can be used for solving such navigation problems. One method is to use a hierarchy for planning. In a hierarchical system the main roads might be described on a coarse scale map where a plan for the broad outline of getting from A to B could be worked out. Then, using fine scale local maps, the route could be examined in detail to produce the final plan. Another method, known as *divide and conquer*, relies on finding a mid-point or several intermediate points which can be considered as way-stations on the route to B. If the route is constrained to pass through such points it reduces the planning problem into a series of smaller planning problems. This is the process of decomposing the original goal into a series of sub-goals.

These route-planning examples illustrate a fundamental principle: any available domain knowledge should always be exploited wherever possible. Blind search will invariably produce very poor performance, but, by being augmented by additional information, an informed search is created which offers much greater potential. Unfortunately, domain knowledge is often not easy to capture or incorporate into search routines.

Example 2 — Game playing systems*
Games such as chess, go, and drafts (or checkers) are ideal games for computerization. This is partly because they are games of perfect information as all the pieces are visible to both

Machinery for thinking about actions

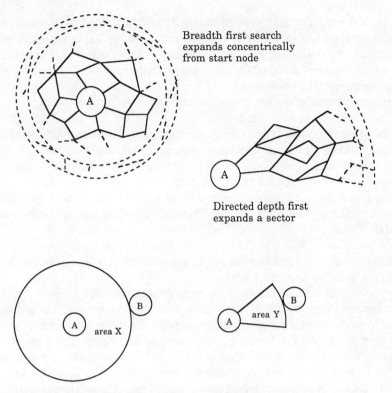

Figure 6.9 Search patterns in a grid of nodes

players and there is no element of chance or probability. This means that it should be possible to calculate the ideal move for any given situation, and hence, in theory at least, a computer should be able to play perfect chess. We will describe a few features of game-playing systems here and then examine the reasons why their promise is not always realized.

The general problem is: given a board state and the rules of the game, what is the best move that should be made? The rules describe legal moves that can be made, and so define operators which transform states in a game tree. Each level in the tree alternates between players. First, the computer may make a move, then the human opponent, and then the computer, and so on. Each move is an OR choice, and so the state space forms an OR tree and the basic methods of state space searching can be used. The standard technique is to generate successors of the board state, by trying all possible moves, then evaluate the successors with a special purpose evaluation function to decide their worth and choose the best one. Of course, if this is only done for one level then the machine will appear pretty

---

*These are not strictly relevant to our interests but are included because game playing has been the context for many search developments. If they are skipped by readers, the main points, on p. 112 should be consulted.

stupid, as it will not see wins and good moves that lie just ahead of it. So the process is performed recursively; at each level the search routine is applied to all nodes to examine their successors. At some stage the computational resources will be exhausted and so the depth of search must be limited. (Due to the combinatorial explosion there is absolutely no hope of going through the whole tree.)

The main components of a game playing system are:

1. A generation routine which finds successor states through the implicit game tree. This procedure includes termination mechanisms for maximum depth limits.
2. An evaluation function, designed and implemented for the given game. This is used in deciding the worth of particular board positions.
3. A backing-up technique is used in order that the value of nodes found deep down in the tree can be reflected higher up the tree. This enables a final selection to be made at the root of the tree.

The whole process is naturally recursive as the state space search trees for these games are essentially recursive data structures.

Taking each of the three components in turn, the generation routine can be any search technique but is usually depth first. The reason for this is that breadth first and its variants are unusable as the combinatorial explosion involves far too many nodes to store. Depth first can do a good job of examining nodes down to a certain depth without excessive memory overheads. In fact, as we shall see later, depth first is also a good method for augmenting with domain knowledge about the state of the game. Secondly, an evaluation function is a routine or procedure which takes as input a particular board position and gives a numeric output that reflects the value of the position for a player. So, for example, the positions shown in Figure 6.10 are respectively, good, bad and indifferent for the cross player. The computed values represent the relative worth of the positions. The way these values are calculated depends upon the type of game and will vary with the stage of the game and how near or far from a winning position it is. So the evaluation function is *very* application-dependent; it uses very specific domain knowledge. Also it *must* be very fast to compute because it is going to be used many times to evaluate nodes during *every* single search for a move. There is no reason why evaluation functions shouldn't give more than one value, but they nearly always have this single dimensional property because of the requirements for speed and efficiency. In real life, we would take many factors into consideration when evaluating the value of a board or position, and this, in fact, is often done inside evaluation functions, but the final result is usually a single value for comparison purposes. In chess playing, the evaluation result is usually created from a polynomial in which the terms of the polynomial are various features, such as the number of pieces on the board for each side, the control of the centre of the board, the advantage in dynamic movement for various sides, and so on. Then these features are combined with weighting and summing functions to produce an overall figure of merit. Of course this summation of features may well be a serious shortcoming as detailed information about the various aspects has been lost after the evaluation. Nevertheless, the complexity of the problem of trying to evaluate multi-dimensional measurements makes it very hard to incorporate effectively into search methods and is not yet well enough understood.

The third feature of search-based game playing systems is the backing-up method. This is

Machinery for thinking about actions

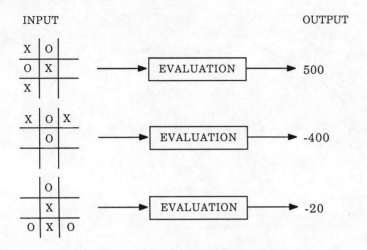

Positive values are better for X and worse for O

Figure 6.10 An evaluation function for a simple game

a technique by which the values of nodes at the terminals of the tree are passed upwards so that a move can be chosen from the top. In game-playing of the two-person zero-sum type, e.g. chess, one person will be trying to maximize the evaluation function results while the other player aims to minimize them. The backing-up process involves applying the evaluation function to the terminal nodes, then propagating alternate maximum and minimum values upwards through the tree. Finally, at the root of the tree the best value will determine the move that is recommended. This technique is known as the *mini-max method*. It assumes optimum play by both sides. Assuming that both players do play as well as they can, this method guarantees to produce the best possible result for the given depth. Figure 6.11 shows an example of a search tree with the terminal nodes drawn in, the backing-up values can be seen for each level, and the final choice is made on the basis of these from the top node. The final result leads to the best node that could be expected when playing against an adversary; it does not lead to the best value node in the tree. Notice that the mini-max values can be computed and passed back *while* the tree is being generated in depth first manner, thus avoiding the need to store the whole tree. This is one of the advantages of depth first searching.

Nearly all chess-playing programs and many other game-playing programs use these basic techniques, that is, depth first search, specific evaluation of terminal nodes, and mini-max backing-up. Many variations are possible but the basic technique seems to be universal. There are many improvements that have been made to game playing systems. Some of these are purely search efficiency speed-ups and so do not cause any hazards in the sense that important branches of the tree might not be examined. Other methods definitely are hazardous and use heuristics to chop off branches in the hope of ignoring the nodes which would prove uninteresting, while saving resources to explore the more important or interesting ones.

The use of heuristics invariably involves pruning the state space tree in some way, particularly at the higher levels of the tree. These are high-risk techniques but are necessary

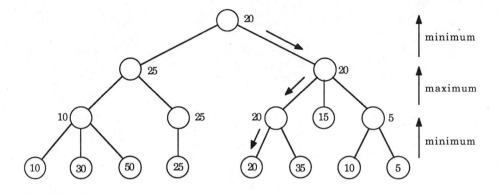

Evaluated nodes are shown with internal values
Backed up nodes have external values

Figure 6.11 Mini-max backing-up of branch evaluation

to provide a good reduction in the amount of tree that remains to be processed. For example, beam search is a hazardous process but has the property that it can be relied on to deliver a fixed number of nodes at each level. This compromise between thorough searching and practical performance is forced on us by the combinatorial explosion.

There is another trade-off involved in heuristic searching — the relationship between the quality of the heuristic data and the need to search deeply for good results. If our board state evaluation function is rather primitive, we can back up the results from deeper levels to improve their reliability. Theory tells us that backed-up values for a node are more accurate than the direct evaluation of the node. (This is the reason why many game-playing systems have fast but crude evaluation functions and obtain good results through deep back-ups.) Conversely, if the evaluation method was extremely accurate, we wouldn't need to search at all as an evaluation of the successors would immediately identify the best move to make! In practice, the skills of the designer, particularly in capturing and effectively employing good quality heuristics, will largely determine the performance of any search system.

Key points that emerge out of game-playing studies are:

1 The great value of application-based heuristics to provide guidance for an informed search.
2 The futility of attempting any kind of blind search on a combinatorial growth problem.
3 The importance of evaluation functions to provide estimates and heuristic data.
4 The paradoxical balance between the quality of heuristic knowledge and the thoroughness and depth of search.
5 A possible view of problem solving as a game in which the environment is seen as an opponent.

Example 3 — Blocks world planning
Perhaps a more relevant example for robotics is the problem of planning a sequence of actions so that a given world configuration is changed into a new configuration. In order to

Machinery for thinking about actions 113

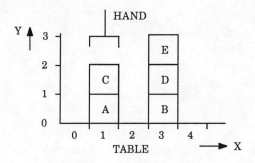

Figure 6.12 The blocks world

develop some simple programs to illustrate these ideas we will use a very simplified two-dimensional blocks world model. Figure 6.12 shows this world, which contains only one type of block. The two-dimensional world is labelled with integer coordinates and blocks are only allowed to occupy unit squares in this world. Blocks with consecutive values are considered adjacent. A robot gripper moves in the same coordinate system and can pick up a single block provided it is above the block. This GRASP operation is one of the fundamental actions available; another is MOVE-HAND (from one coordinate position to another). The blocks are labelled A through to Z and each block has associated data containing entries for its position, what is below it, and what it supports. The gripper is called HAND, and also has maintained data regarding its position and what it is grasping. Another variable is called TABLE and contains the names of all the blocks which rest on the base line, that is, all blocks with Y coordinates of 0. We recognize the problem of planning a new configuration is the same as the task of finding a particular path through the space of possible actions. This assumes that the actions are treated as operators which can change a state of the world into new states and also assumes that the task is defined in terms of start states and goal states. The planning problem then becomes one of searching for an acceptable path, not necessarily the optimum path, to reach the goal state. When a suitable path has been found, this path can be output as the plan of action that is required to achieved that goal. Hence, we can see how an automatic planner could be based on a searching system.

We now specify an operator, called PUT-ON, that can be defined in terms of preconditions and state transformations. The operator brings about the changes that would occur if a series of GRASP and MOVE actions were used to place one given block, x, onto another block, y. Figure 6.13 shows the operator and the world data base. The state of the world is represented by a set of factual statements in the data base, e.g. on(A B), clear(A), etc. Notice that the geometric values needed for robot instructions can be derived from data on movements and block positions. The necessary coordinate changes are easily incorporated into the operator, but for reasons of clarity they are not shown in the database or in the workings of this planner.

The operator essentially states that if x is to be placed on y, then the top of x and the top of y must be clear and x cannot be the table — those are the preconditions. The transformation part, i.e. the result of this operator, is that x is moved from its previous place

```
( on C A )
( on A TABLE )
( on E D )
( on D B )
( on B TABLE )
( clear C )
( clear E )
```

Operator   (put-on x y)
Preconditions   (clear x) & (clear Y)   [x ≠ TABLE]
Actions on database
        (on x z) is retracted
        (on x y) is asserted
        (clear z) is asserted   [if z ≠ TABLE]
        (clear y) is retracted

Figure 6.13   Database and operator for blocks world

to y, the previous place is considered to be now clear (if it was not the table), and the block, y, on which x has been placed is no longer clear. If we define only this single operator (in order to simplify the illustration), then our search must find a suitable sequence of such PUT-ON operations to achieve the goal. Other search systems may employ several lower-level operators of different types such as MOVE-HAND, GRASP, MOVE-OBJECT and UNGRASP, but our PUT-ON is essentially equivalent to a subroutine built from a series of these. In any case, the principle is the same — we have to find a series of operator applications to transform the start state into the goal state.

Now suppose we wish to write a search program which will generate simple plans, given various operators, start states and goal states. Our first attempt might be an exhaustive blind search that sequentially tries all the possibilities in turn, and we may well decide that the depth first method is the most appropriate because (a) it is very easy to program and implement, (b) it will eventually search all the tree down to a particular depth, and (c) we can control the depth of search easily. The search begins at the start state, and one of the possible instantiations for variables x and y is generated for PUT-ON. This leads to a new state and further operator applications are performed. Eventually, either the goal state is reached or the maximum depth limit initiates back-tracking through other branches of the state space. The search process ensures that all valid combinations of PUT-ON are pursued down to a given depth limit. Some configurations are shown in Figure 6.14 for two start states and a desired goal state. The start state, called EASY, only requires two PUT-ON actions to achieve the goal, whereas start state HARD requires seven applications of PUT-ON because block A must be removed from the bottom of the pile. In the author's experimental program, there is a parameter for the maximum depth limit — let's call this EFFORT. A test has been incorporated to prevent the last most recently moved block from being moved in consecutive operations. If this is not prevented rather neurotic behaviour can be produced by the planner — blocks get placed on top of other ones and then immediately removed. The basic structure of the search program is shown on page 115.

Machinery for thinking about actions

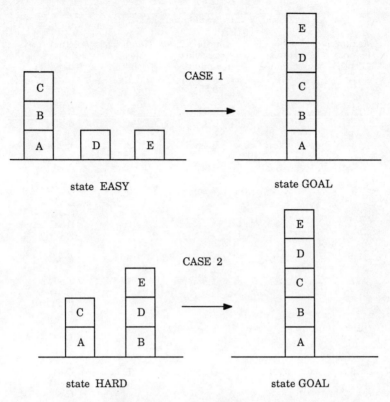

Figure 6.14 Two searching tasks

Blocks World Search Strategy

Function SEARCH (Current-State, Goal-State, Effort, Last)
 — if Current-State = Goal-State then exit (success)
 — if Effort = 0 then exit (failure)
 — generate Successor-list from Current-State (excluding Last) (by applying PUT-ON to all valid block combinations)
 — for each in Successor-list — assert as new Current-State
  — decrement Effort
  — record Last
  — call SEARCH (recursively)

The program was tested on a series of trials using both start states and with maximum depth values (EFFORT) set from 1 to 8 inclusive. A summary of the results is given in Figure 6.15. The results for start state EASY show a common problem with depth first searching. Notice

Case 1 - starting at state EASY

Maximum search depth	Number of states examined	cpu time s = seconds m = minutes	Number of steps in solution	Actual solution steps
1	7	0.55s	-	none
2	28	1.9s	2	D->C E->D
3	153	9.4s	2	as above
4	795	44s	2	as above
5	2971	2.7m	5	C->TABLE D->E C->B D->C E->D
6	13164	12m	5	as above
7	2089	1.9m	7	C->E B->D C->TABLE B->A C->B D->C E->D
8	9566	9.9m	8	C->E B->D C->TABLE B->A D->E C->B D->C E->D

Figure 6.15 Results for search based planner

that *whatever* effort level is chosen a solution is always found at or near that maximum depth. This close correlation between EFFORT and the number of steps in the solution indicates that there are many repetitions of the goal state throughout the state space. In this case there are many solutions to our problem because many different branches of the tree converge onto the right configuration of blocks. Depth first search acts as though it were processing through a tree structure (where each node has only one path to it from the root), whereas the repeated states indicate that it is a cyclic graph that is actually being searched. This is the major shortcoming of our search process: because repeated states are not detected our search tree is potentially infinite. This planner could keep building up configurations and knocking them down, *ad infinitum*. Consequently, if the search is allowed to descend to depth 5 or 8 then it is likely to find a solution at that depth before it is forced to back-track and find less deep solutions. Thus, in this particular problem, depth first searching not only finds poor solutions but also takes longer to find them if allowed to do so!

Case 2 - starting at state HARD

Maximum search depth	Number of states examined	cpu time s = seconds m = minutes	Number of steps in solution	Actual solution steps
1	4	0.27s	-	none
2	18	1.0s	-	none
3	85	4.9s	-	none
4	433	24.8s	-	none
5	2204	2m	-	none
6	10297	9.4m	-	none
7	8080	7.3m	7	C->TABLE E->TABLE D->E B->A C->B D->C E->D
8	10737	11.2m	8	C->TABLE E->C D->TABLE B->A E->TABLE C->B D->C E->D

Figure 6.15 Results for search based planner (part 2)

Other techniques could improve on this considerably. Breadth first, for example, would be much superior. Some experiments were run with breadth first search and they always returned a result for level 2. However, in general, breadth first is not possible for deeper solutions because of its enormous storage overheads. The use of a *best first technique* would solve most of these problems by maintaining a set of unique, best-yet states in the check lists. But this now introduces special knowledge because *best* must be defined in terms of the application problem. One node is better than another when it is nearer to the goal state, but how do we determine when we are getting nearer to the goal? We need to find some measure of 'nearness-to-goal'. As soon as we start to think of these modifications, we are moving away from blind searching towards informed search procedures that have information about the domain they are working in. In fact, if very much information is introduced in this way, we may find that back-tracking is not even necessary. For example, with a really good measurement of nearness-to-goal we could try a simple hill climbing algorithm, such as make any random move that increases the nearness-to-goal value.

Another example of the use of domain knowledge is the obvious heuristic: un-stack all the blocks on to the table, and then build up the configuration from there, starting at the lowest level block. This is almost certain to work, but only because it is based on prior knowledge of block stacking problems. There is no way that the computer could discover such methods on its own without much more knowledge of the world of blocks.

There are other things to note from the results. The solution for the HARD state at level 7 takes less time (less states are examined), than the full examination of the state space for the failure at level 6. So some cases might actually need less work to explore deeper paths in the tree. Also, notice that HARD generates fewer solutions than EASY in the early stages because there are less *free* blocks in HARD than in EASY and hence there are less options for operator applications. If there is only one clear top, then there is only one action possible, i.e. move that free block. This helps to explain why an apparently simple problem, with many options in the early stages, can be quite difficult to solve. We must capitalize on any available information that indicates which are the interesting avenues to explore. With blind search every avenue is just as interesting as any other and therefore it is highly likely that it will be very inefficient in finding a solution.

These experiments give us some idea of the shape of the state space, how the states relate to one another and the numbers of states at different levels. They also indicate that blind searching is really only useful as a last resort when no other data are available. It is quite likely in any robotics application that more data *will* be available and there will often be known priorities for various options. Such data introduce constraints that rapidly cut down the magnitude of the searching problem. This will be discussed in later chapters. However, as in game playing systems, an effective informed search can be created, even when exact knowledge regarding the solution is not known, by using heuristic rules-of-thumb to guide the search in roughly the right direction. As a general guideline (a meta-heuristic!) the amount of searching required varies inversely proportional to the quality of the heuristics. With very high-quality heuristics one hardly needs to search at all, because the information directly points the way to a solution. The difficulty is that all introduced heuristics will be very application-dependent and have to be designed and implemented for each different task. As this discourages general-purpose designs it can be very wasteful because there is little carry-over of principles or methods between different problem areas. Notice also that sensory data have not been included in our searching strategies at all. This seems rather silly, in terms of human experience, as sensing can immediately provide data for the correction or compensation of mistakes in a plan. Sensory data can also help to resolve decisions when different options appear. Clearly, the incorporation of sensory data into searching methods is worthy of much investigation.

For these reasons it seems that search is a very useful technique to support other systems, but it seems unlikely that it would form the core of a reasoning system that would satisfy our requirements. Search can be viewed as one of the tools in our problem-solving armoury but is not nowadays viewed as the ultimate panacea.

6.3 Goal-directed planning

There is a blocks world planning technique which seems to have many desirable features for our planning requirement. This approach has similarities with the recent programming methodology known as *object oriented programming* but we describe here the earlier ideas that have emerged over several years, mainly out of robotics work done at MIT. The basic idea behind this approach is based on collections of goal-directed procedures. These are short procedures that perform a particular goal or sub-goal and have access to a common

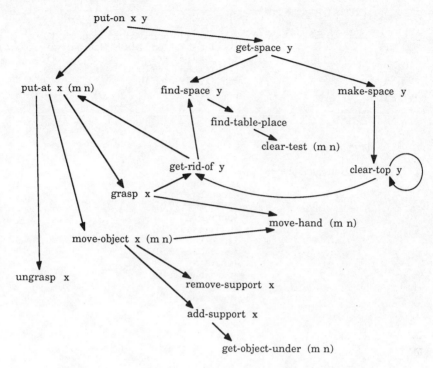

Figure 6.16 Relationships between goal-directed procedures

database. Each procedure is aimed at some individual goal and achieves this by breaking up its task into a series of sub-goals. Each procedure is designed to be self-contained so that it may be called at any stage and should be able to encounter any situation without performing erroneously. There is no overall executive control; each procedure calls others as sub-goals and the flow of control in the system depends on the data-directed calls which are produced by current planning needs. Figure 6.16 shows the way in which the procedures interact and it is clear that this is a mutually recursive system and quite complex sequences of procedure calls may occur.

There are some primitive routines which the planner can call as the lowest-level basic sub-goals; for example, a function called CLEAR-TEST will examine a given location and indicate if it is occupied. The procedures GRASP, UNGRASP and MOVE-HAND are robot actions, which, when called, add a note to the plan to record the next required action step. In order to operate the system, a desired goal is called. Usually this will be a top-level goal like PUT-ON A B, which states that the goal of A placed on top of B is to be achieved. Frequently this cannot be achieved directly, as A and B may have blocks on top of themselves and, in order for the hand to pick up a block, its top must be clear. Thus, the planner will break down the PUT-ON operation into a series of CLEAR-TOP, MOVE, and GET-RID-OF operations until the final plan of achieving A on B has been reached. In Figure 6.17 before and after diagrams for a particular put-on goal are shown. In this system the blocks world database is very similar to the search example but does not maintain data on the blocks with clear tops. Instead,

INITIAL STATE

```
2|      E
1|  C   D
0|_ A___B___
   0  1  2  3
```

A - (at (1 0)) (on TABLE) (supports C)
B - (at (3 0)) (on TABLE) (supports D)
C - (at (1 1)) (on A)
D - (at (3 1)) (on B)   (supports E)
E - (at (3 2)) (on D)

INPUT GOAL: put-on A D

OUTPUT

The generated plan is:

(move-hand (3 2))
(grasp E)
(move-hand (0 0))
(ungrasp E)
(move-hand (1 1))
(grasp C)
(move-hand (2 0))
(ungrasp C)
(move-hand (1 0))
(grasp A)
(move-hand (3 2))
(ungrasp A)

FINAL STATE

```
2|      A
1|      D
0|E   C B___
   0  1  2  3
```

A - (at (3 2)) (on D)
B - (at (3 0)) (on TABLE) (supports D)
C - (at (2 0)) (on TABLE)
D - (at (3 1)) (on B)   (supports A)
E - (at (0 0)) (on TABLE)

Figure 6.17 A simple planning task for the goal based planner

support relations are stored and the system can then work out which blocks are free to move. Also, coordinate data for each object are held in the database and these are manipulated directly by the lower level procedures. The output plan list is also shown and the small steps involved in moving the blocks are clearly evident in the plan. Figure 6.18 shows a trace of the system operating on this problem and indicates how procedures may call many others to achieve the sub-goals. The operation is rather subtle; this trace only shows the names and parameters of the procedures when they are called, and this order is different from the plan order because plan steps are generated towards the *end* of procedures. This is a very attractive style of planning as the procedures can be written separately for each goal that must be achieved for each simple action. However, there are problems because this methodology assumes that each procedure is completely independent and cannot be disturbed by the operation of others in trying to achieve its sub-goals. This, in general, is not easy to prove for any given situation. It is also difficult to program because it is very hard to anticipate all of the different ways in which the system may react. The system may work very well in achieving certain goals, but at a later stage, an erroneous planning process might be found which had lain dormant during all previous testing. This

Machinery for thinking about actions 121

        Plan steps are generated by procedures marked **bold**.
        Indentation shows procedure calling hierarchy.

```
put-on A D            (Initial Input Goal)
  get-space D
    find-space D
    make-space D
      clear-top D
        clear-top E
        get-rid-of E
          find-space TABLE
            find-table-place
              clear-test (0 0)
          put-at E (0 0)        (Remove E from top of D)
            grasp E
              move-hand (3 2)
            move-object E (0 0)
              remove-support E
              move-hand (0 0)
              add-support E (0 0)
                get-object-under (0 0)
            ungrasp E
  put-at A (3 2)                (Move A to target location)
    grasp A
      clear-top A
        clear-top C
        get-rid-of C
          find-space TABLE
            find-table-place
              clear-test (0 0)
              clear-test (1 0)
              clear-test (2 0)
          put-at C (2 0)        (Remove C from top of A)
            grasp c
              move-hand (1 1)
            move-object C (2 0)
              remove-support C
              move-hand (2 0)
              add-support C (2 0)
                get-object-under (2 0)
            ungrasp C
      move-hand (1 0)
    move-object A (3 2)         (Complete move of A)
      remove-support A
      move-hand (3 2)
      add-support A (3 2)
        get-object-under (3 2)
    ungrasp A
```

Figure 6.18   Trace of goal directed planning program

style of planning can be seen as analogous to an associative network of procedural data. One can view the goal procedures as nodes in a data-driven network in which different nodes will be active at different times, thus producing a sort of dynamic procedural net.

While these procedures can be very effective and appear quite straight-forward in simple examples, there can be serious difficulties from a software engineering viewpoint when a large system is being built consisting of many thousands of lines of code. The procedures should be short, goal-oriented and self-contained. They must also compute all their relevant

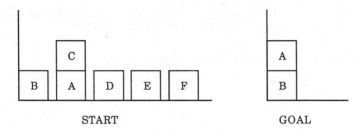

Figure 6.19   Goal conflicts

pre-conditions before invoking their sub-goal actions. However, while they execute, they do not know the partial states or results of the other procedures, and can only communicate through the common database. If they require intermediate results that have not been placed in that database, then they are effectively ignorant of situations which may be important to them. Clearly, this design will work very well if complete independence of procedures can be guaranteed. But if they interact in any non-trivial sense, it can be quite difficult for designers to define suitable interfaces between the procedures.

The recent development of *object oriented programming* offers a more rigorous approach to the problem of organizing a set of interacting processes. In this method, procedures (known as 'objects') can only communicate by passing formal messages to other objects. This provides structuring constraints that the above method lacks. Other features include the ability to instantiate objects from general class descriptions. However, no amount of structuring effort will be worthwhile if the problem can not be rendered into smaller parts, due to either intrinsic complexity or because of bad presentation. In other words, not only must the task be decomposable, but we must also be able to *find* a suitable decomposition in order to map it onto a set of goal-directed procedures. This is very difficult to guarantee in general. Consequently, we cannot assume independence and some additional mechanism must be incorporated into goal-directed planning systems.

One difficulty with goal-directed programming is that only one 'on' goal is achieved at once. In the search example, a single goal state could define several 'on' relations. If we call PUT-ON several times to achieve a multiple final goal state it is quite possible for one execution sequence to disturb or destroy a goal that has already been achieved. One method of dealing with this is to introduce *protection mechanisms*. These ensure that once a goal has been achieved it will be protected while other parts of the plan are completed. This requires special labels for the protected items and additional methods for maintaining lists of currently protected sub-goals. Goals which have been achieved can then be recorded and held fixed and any further operations by other procedures will not disturb these partial achievements. Figure 6.19 gives an example of this. If in trying to achieve A on B, the planning system removes block C, and there is no room on the table, it may be placed on top of B (because the FIND-TABLE-PLACE procedure examines the left-most locations first), and this would interfere with the original goal of A on B. In this case, protecting the clear-top of B, while the object is removed from A, would solve the problem because C would then have to find another position. Unfortunately, this leads on to another problem, because it is not easy to find out which sub-goals must be protected. If we protect too many, we may provide so many constraints, that the problem cannot be solved. So we must find a small number of critical goals to protect; and yet we have no criteria for deciding which are the important ones.

This problem closely relates to the frame of reference problem. The difficulty with automated reasoning is that any action might change some of the relations in the world, but there has to be some method of knowing which ones have been altered. If a planner has no knowledge of the consequences of each action, then it must look at *all* of the facts in the world after every action, to see if anything has been altered. One technique, which became known as the *frame axioms method*, tried to get round this problem by listing all the things that were not altered after an action. However, this may be suitable for theorem-proving systems, but in the real world most actions do not change very much of the world. So it is much more usual to follow the idea that all relations are assumed unchanged unless explicitly mentioned in an action definition. Thus, an action like MOVE-OBJECT A (1 3) should have some information attached to it that says that this action might disturb the state of other blocks that are near A or otherwise have some relation with it. Such frame conditions are easily added into the descriptions of the operators if the world is reasonably independent and not many actions change many of the facts. (It is easy to demonstrate these ideas in our simple blocks world.) Difficulties occur when complicated actions, often called *influential actions*, effectively change some of the facts through indirect routes. An example of this would be the case when a table or tray is moved. This action only appears to affect the table, but any objects on top of the table will also have their positions changed, and so all their relationships to the rest of the world will be altered. If the table move action only records details about changes to the table, this becomes an 'influential' action with indirect side effects. The problem now is that the interactions become very complex and we must check all the items on the table to see if they would fall off if they would hit other objects, or if any other effects are likely to occur. Hence, the more influential actions there are, the more computation there will be associated with each action to deal with the updating of the world model.

6.4 Rule-based planning

Following our examination of production systems in Chapter 5, we note that production rules could also form the basis of a planning system in the blocks world. Production rules consist of two parts, the condition part and the action part, and we can fairly easily cast our blocks world planning problem into a series of such rules that express basic actions in the world. Figure 6.20 shows a typical set of rules for this problem. For example, Rule 2 says that if some block, x, is to be placed on the table, and x has a clear top, then we can remove the fact that x is on a previous block, y, and inform the system that y is now clear, by asserting (clear ?y). Then we can assert that x is now on the table and erase the action request which has now been satisfied. All these assertions are stored in a global database, called WORKING MEMORY, just as for the other programs. This type of system has similarities with the AI programming language PROLOG and, in fact, it is very easy to write a production interpreter for such a scheme in the PROLOG language. This has been done, and a trace of the production rules working on the same problem as for the goal based planner is shown in Figure 6.21.

--- rule 1 ---
```
(goal ?x ?y) & (on ?x ?y) &
(goal ?p ?q) & (on ?p ?q) &
(goal ?r ?s) & (on ?r ?s)    --->   write ( RULE 1 - GOALS SATISFIED )
                                    stop
```
--- rule 2 ---
```
(put-on ?x TABLE) & (clear ?x) --->   write ( RULE 2 - MOVE ?x TO TABLE )
                                      erase (on ?x ?y)
                                      assert (clear ?y)
                                      assert (on ?x TABLE )
                                      erase (put-on ?x TABLE )
```
--- rule 3 ---
```
(put-on ?x TABLE ) & (on ?y ?x) --->
              write ( RULE 3 - CLEAR ?y FROM TOP OF SOURCE ?x)
              assert ( put-on ?y  TABLE )
```
--- rule 4 ---
```
(put-on ?x ?y) & (clear ?x) & (clear ?y) --->
              write ( RULE 4 - MOVE ?x TO TOP OF BLOCK ?y)
              erase (on ?x ?z)
              assert (on ?x ?y)
              erase (clear ?y)
              erase (put-on ?x ?y)
```
--- rule 5 ---
```
( put-on ?x ?y) & (on ?z ?x) --->
              write ( RULE 5 - CLEAR ?z FROM TOP OF SOURCE ?x )
              assert (put-on ?z  TABLE )
```
--- rule 6 ---
```
(put-on ?x ?y) & (on ?z ?y) --->
              write ( RULE 6 - CLEAR ?z FROM TOP OF TARGET ?y )
              assert (put-on ?z  TABLE )
```
--- rule 7 ---
```
(goal ?x ?y) & (on ?x ?z) --->
              write ( RULE 7 - CREATE MOVE FOR GOAL ?x ON ?y)
              assert (put-on ?x ?y)
```

Figure 6.20   Production rules for block tasks

The data in WORKING MEMORY are in the same form as for the search planner and object coordinates have not been included in the rules. The only rather unusual feature here is that desired actions, e.g. (PUT-ON A B), are held in WORKING MEMORY as declarative requests together with all the other facts about the world. So instructions are just seen as other facts and we must remember to add and remove them from the WORKING MEMORY as required by

Machinery for thinking about actions 125

Output from production system:

    RULE 7 - CREATE MOVE FOR GOAL **A** ON **D**

    RULE 5 - CLEAR **C** FROM TOP OF SOURCE **A**

    RULE 2 - MOVE C TO TABLE

    RULE 6 - CLEAR **E** FROM TOP OF TARGET **D**

    RULE 2 - MOVE E TO TABLE

    RULE 4 - MOVE A TO TOP OF BLOCK D

    RULE 1 - GOALS SATISFIED

Only rules 2 and 4 are able to change the physical world by calling for robot action. Executions of these rules are shown **bold**.

WORKING MEMORY CONTENTS

Before	After	
(clear C)	(clear C)	
(clear E)	(clear E)	
(on C A)		------ deleted fact
(on A TABLE)		------ deleted fact
(on E D)		------ deleted fact
(on D B)	(on D B)	
(on B TABLE)	(on B TABLE)	
(goal A D)	(goal A D)	
(goal D B)	(goal D B)	
(goal B TABLE)	(goal B TABLE)	
	(clear A)	------ new fact
	(on C TABLE)	------ new fact
	(on E TABLE)	------ new fact
	(on A D)	------ new fact

Figure 6.21   Trace of production rule planner

the various rules. These action requests are placed in WORKING MEMORY by rules that desire the action and are removed by rules that perform the action. In this example, action requests, i.e. PUT-ON, can be generated by rules that need various blocks to be moved (rules 3, 5, 6 and 7). Rules 2 and 4 are the rules that actually move blocks, and this is achieved by writing out suitable messages. The plan steps are thus generated by recording the output of Rules 2 and 4. These two rules perform updates on WORKING MEMORY to reflect the changes caused by the moves and they also remove the action requests as they are completed. Consequently, no PUT-ON statements are seen in the before and after diagrams of Figure 6.21 as they only exist while the system operates.

We also must invent a statement to describe what is to be achieved by the system. This is the goal statement, e.g. (goal A B) which means that A is required to be on top of B. Goals are matched with on statements, as in the first rule, so that if the goal is A on B and there is also a *fact* in WORKING MEMORY which states that A is on B, then we can write out a message that the goal has been achieved.

Because of the way in which a production system operates it can be quite difficult to detect bugs in the rules. For example, the omission of one minor condition error in one of the rules, might well cause a completely different sequence of actions to occur. Another problem is that different interpreters will use different methods of conflict resolution to choose between several matching rules. In the system shown, the interpreter always chooses the lowest numbered matching rule to fire. This knowledge is important when deciding the order the rules are entered. In Figure 6.20 the goal test is given first, then follow rules that deal with special cases (using the named object — TABLE), then comes the general PUT-ON operation followed by cases where it is blocked, and finally there is a rule that generates movement requests.

The subtlety of production systems can be seen in the design of the rules shown in Figure 6.20. Rule 3 might seem unnecessary as it looks like a special case of the more general Rule 5. Rule 3 says 'if x is to be placed on the table but something is on top of it then try to place that something on the table'. This is exactly the same as Rule 5 if $?y$ is bound to TABLE. However, if we run the system without Rule 3 we find that Rule 5 can get into an endless loop and fill the WORKING MEMORY with copies of the same action request. This happens when a lower block in a pile is to be moved: first, Rule 5 requests a block to be placed on the table, but then Rule 2 is unable to do this, as there are more blocks on top. Thus, Rule 5 matches again and asserts another request and so on ....

Our rule-based planner has many similarities with the goal-directed technique, but it has a more uniform design. Notice that the production system shown is a much more straightforward version of the goal-directed procedures which are only indicated in Figure 6.16. In the goal-directed case, the procedures are fixed in their calling sequence by the procedural code. In production systems the rules may interact in a very flexible and fluid way, and so they represent a more open and versatile level of control than the goal-directed procedures.

There are also other advantages with production systems in that the architecture is open and uniform. This means it is an ideal environment for experiments on many AI ideas. Not only can the rules be easily changed but the control of the interpreter cycle is also accessible. Different methods of conflict resolution and matching schemes can be investigated with much more ease than with many other styles of programming. In addition, production systems offer a framework for learning systems. By writing rules that assert into the production memory it is possible to synthezise new rules. Thus, it is then possible to produce a general rule which captures the information contained in several other clusters of rules.

6.5 Blackboard systems

There is yet another version of distributed control systems with an even broader architecture — this is the *blackboard system*. Blackboard systems contain many separate procedures, known as *knowledge sources*, which independently process knowledge and communicate through a global database known as the 'blackboard'. This is a particularly appropriate design when there are different *styles* of knowledge available and yet the knowledge interacts in various ways. Figure 6.22 shows a schematic diagram of an example system in order to illustrate the main concepts. In this example, we have one program that deals with the geometric features of a robot world, another system knows about the ways in which

Machinery for thinking about actions

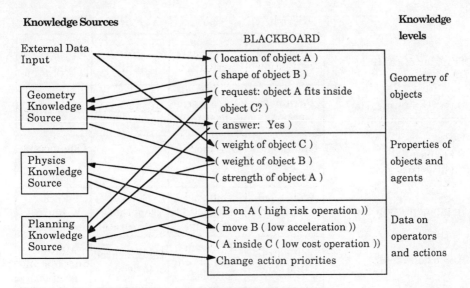

Figure 6.22 Blackboard system architecture

physical variables (such as weight and rigidity) affect the execution of actions in the world, while yet another module is a knowledge-based expert that knows how operational parameters affect different plans of action. Clearly, these different aspects of the same problem will need to interact, and the blackboard scheme provides a useful style of architecture to allow this.

Each knowledge source can operate independently and concurrently. As they are autonomous entities some might be implemented as groups of rules, others might be procedures or even large programs, and others could be direct or processed input data from sensors or external sources. During operation, knowledge is placed on and removed from the blackboard in a constantly changing flow. Notice that knowledge sources can operate both within and across the boundaries between knowledge levels. In the next chapter a more detailed application of blackboard systems is given.

Blackboard systems are a further step on the broadening of the spectrum towards distributed control. There are similarities with production systems in the open communication area, the way modules capture control, and the pattern-directed or data-driven processing style. The main difference is that the unified rule structure has now been replaced by heterogeneous processing modules. This loses the advantages of a uniform notation but gains benefits by combining the power of disparate sources of knowledge. Such a design of system is very suitable for dynamic problems of interaction and interpretation, for example in sensory data processing where many different signals are arriving from different sources and need to be coordinated. But, as indicated in Figure 6.22, blackboard systems are also appropriate in robot and other planning situations. They seem very suitable for sensor-based planning and for situations where there is a need to monitor a robot plan while it is being executed.

	Search	Goals	Rules	Blackboards
Transfer of control	Recursive activation	Sub-goal activation	By capture	By capture
Data communication between modules	Parameter passing	Parameter passing	Broadcast	Broadcast
Fit to parallel architecture	One module per recursion	One module per sub-goal	?	One module per K-source
Implementation factors	Well-tried, easy method	Sub-goal interaction problems	Simple, separate interpreter	Each K-source different
Execution factors	Difficult to fine tune	Program fixed for one problem	Complicated to debug	Multiple, incremental solutions

Figure 6.23 Features of knowledge processing schemes

6.6 Summary

In the four systems we have seen for planning actions in the blocks world, there seem to be a range of options which extend along a spectrum from the narrow and precise to the broad and general. Figure 6.23 shows a table of some characteristic features. In particular, the method of control is quite different between search-based planning and rule-based planning. In search, first one option and then the next option is chosen in a fairly mechanical manner. In rule-based planning, it is not obvious which option will be chosen next, as this depends upon the prevailing data at the time. If we imagine the rules to be mini-procedures we could say that they capture control by establishing a match with the WORKING MEMORY, and then execute their action part. Another way of looking at these systems is the way in which their data are passed as messages. In search techniques, data and results are passed backwards and forwards through parameters to particular calling procedures, but in rule-based systems and blackboard systems, the results are broadcast to all procedures, any of which might then pick them up. Another feature is the degree of parallelism that is possible. In blackboard systems the knowledge sources are quite independent and could run as concurrent modules; therefore the scheme is quite suitable for parallel implementation. To some extent, goal-directed systems may also be viewed as parallel systems, as a set of sub-goals could be satisfied concurrently. However, production systems and search methods examine one option at a time so there is little scope for parallelism.

In some sense, all these methods can be seen as a form of search. In its broadest sense, search is the process of choosing between a range of alternatives, and each of these techniques implements some method of doing that (although sometimes in a very diverse manner). Some AI proponents claim that search is the fundamental problem in AI and all the known methods represent variations on the problem of choosing what to do next. We will not pursue that argument, but we hope to show that, although search may be a major technique, there are so many different aspects and considerations in real-life application areas that the formal classical search technique is highly inappropriate. Many problems can be solved by the simple application of readily available knowledge and related inference processes. Also, as soon as the problem of planning is opened up to include wider issues, such as plan execution monitoring, and tasks, such as diagnosis and interpretation, then again the requirements of the task become quite different. Another problem-solving scheme known as *generate-and-test* then becomes very relevant. Generate-and-test has not been included here, as we will treat it later as a diagnostic tool.

The message comes across, once again, that however many techniques we have in our vast collection of AI ideas, these will not work unless we attend to the *specific* aspects of our particular application problem very carefully. In other words, a small amount of relevant knowledge is much more useful than any amount of automatic searching in a blind or uninformed manner.

1 Reasoning about actions is a planning task. Planning may be performed by many methods, the most basic being search techniques.
2 There are many variants of state space searching. The depth first method is easy to implement and widely used. Best first searching offers many tuning options and will detect repeated nodes.
3 The combinatorial explosion is a fundamental property of many state space searching problems. It can not be avoided, only mitigated.
4 Heuristics are often used to limit or guide a search process. They frequently arise as evaluation functions for estimating the merit of a situation. There is often a trade-off between the quality of heuristic data and the amount of search required.
5 Domain knowledge provides the source of most heuristics.
6 We must take care to distinguish general-purpose heuristics from special case heuristics. Specialized heuristics may produce powerful results but we should not expect them to provide general solutions.
7 Future methods of programming industrial systems will rely more on off-line planning to generate 'how to do' instructions from 'what is required' specifications. They are likely to use a range of different methods, including those given here.

6.7 Further reading material

Research on planning has continued with many developments beyond the state space search strategies described in this chapter. The 'NOAH system', by E. D. Sacerdoti in *A Structure for Plans and Behavior* (Elsevier, 1977), deals with hierarchies of partially-ordered plans, and is able to criticize and improve developing plans.

The map navigation example in this chapter was taken from R. J. Elliott and M. E. Lesk 'Route Finding in Street Maps by Computers and People', in *Proc. AAAI–82*, Pittsburgh, 1982.

An early but classic paper on game playing is 'Some studies in machine learning using the game of checkers II. Recent progress', by A. L. Samuel, *IBM Journal of Research and Development* (1967), volume 2, number 6.

The most comprehensive model of the blocks world was developed by S. E. Fahlman in 'A planning system for robot construction tasks', *Artificial Intelligence* (1974), Volume **5**, Number 1, pp. 1–49.

The goal-directed methodology is described in *Artificial Intelligence* (second edition, 1984), by Patrick Winston (Addison-Wesley).

For more details of robot planning problems, Chapter 9 of *Introduction to Artificial Intelligence*, by E. Charniak and D. McDermott (Addison-Wesley, 1985), gives a good treatment of this area.

A classic application of blackboard systems is given by L. D. Erman *et al.*, in 'The Hearsay-2 Speech-Understanding System: integrating knowledge to resolve uncertainty', *Computing Surveys* (June 1980), volume **12**, number 2, pp. 213–253. A good source of material on blackboard systems is *Blackboard Systems* by R. Englemore and A. Morgan (Addison-Wesley, 1988).

Chapter 7

Speech and language: from mouse to man

You must allow for the imprecision of language. It is a point I cannot make too emphatic. The notion that thought can be perfectly or even adequately expressed in verbal symbols is idiotic.

A. N. Whitehead

In Chapters 3 and 4 we examined the problems associated with giving computers the power to see. In this chapter we will briefly look at speech, hearing and language comprehension. For an example application, we recall the dialogue given in the scenario in Chapter 1 and we now consider the speech requirements that would be needed to achieve such performance. Our aim is to evaluate just how realistic these conversational exchanges with manufacturing machines will be, and to appreciate the nature of the research difficulties involved in producing such speaking, hearing robots. We will not examine detailed techniques for language analysis and processing.

The ability to talk, listen and process language has strong overall similarities with the ability to process and understand images. As for vision, we do not really know how humans internally process speech and language and therefore we have difficulties in building a theory for mechanical processing. In many ways the overall research progress of speech processing has had a similar history of successes and failures to that in vision processing.

One obvious difference is that language is a serial phenomenon, being concerned with sequential orderings of symbols or sounds. Vision, being based on highly parallel sensory processing, seems to be further removed from symbolic information than language. For this reason, knowledge-based techniques become evident at an earlier stage in language research. It is clear that each modality, i.e. each different kind of sensory input channel, has different strengths. For example, vision is a spatial channel; spatial representation and reasoning are much more naturally handled by vision than by language. On the other hand, temporal sequences and time-related events are more easily represented and processed by language. Indeed, there appear to be different styles of thinking in humans; 'spatial' thinkers tend to be more skilled at practical, physical and navigational tasks whereas 'serial' thinkers are better at music, languages and logic. There is even some evidence that these

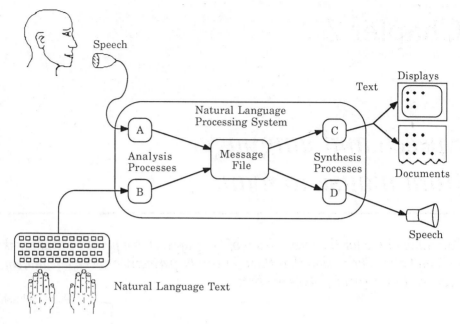

Figure 7.1 Speech and language processing

complementary abilities are specialized in different hemispheres of the brain.

Given that language seems closer to symbolic reasoning, we might expect the understanding problem to be somewhat easier than for vision. However, the general problem of perception turns out to be just as hard!

7.1 The nature of the problem

Figure 7.1 indicates the four different input and output channels that provide natural language communication. For input, we can speak into a microphone or we can type linguistic structures through a keyboard or other character input device. Output can be obtained either as documents (on display screens or on paper), or as speech from a hardware sound-generation system. These four methods of input and output give rise to different ways of using natural language and different types of application.

It is important to notice that it is *natural* language that is being referred to here, i.e. the free-flowing creative messages that are found in any ordinary human communication. These messages are often very idiosyncratic, and can be quite badly formed, both in terms of the rules of good grammar and in the use of words. Also, due to the large number of sources, many messages will be *unique* to the recipient, who cannot have experienced all possible idioms and styles. They should not be confused with the formal and well-defined structures found in programming languages. Programming languages do not have the ambiguities, the flexibility and the infinite variation that natural languages have. Hence, comparing natural languages with programming languages is somewhat like comparing photographs with engineering drawings. In one case, the medium is free and unconstrained and, in the other, there are very strict rules and formalisms which control and constrain the amount of available information.

In Figure 7.1, the message file is a store of message information in an internal form that can be understood, processed and manipulated by the computer system. Almost by definition, the message file must have a well-specified notation to allow computer processing. The sub-systems, A to D, are processing modules which either convert input into the internal notation or generate output. It is clear that the output processes are rather easier to perform as the message is already in a well-understood form. Thus, the conversion into output speech by generating the necessary phonemes and syllables or the generation of natural English documents is *relatively* straightforward, given a good internal notation. Presently, the results may be crude, with grating pronunciation or a boring prose style, but the technology is improving and becoming established. There are examples of this in the audio messages generated by the latest generation of consumer products (e.g. vehicle warning systems) and in the natural language reports generated by some expert systems and other programs which are able to explain their reasoning.

The input processes, A and B, on the other hand, are much more difficult to achieve. This is mainly because the speech and text has to be processed into a form or representation that can be understood internally and yet, in order to control this conversion process, we really need to understand the content: a kind of chicken-and-egg dilemma. This is the reason why we should expect input processes to involve a great deal of interpretation akin to the perception problems mentioned earlier in Chapter 4. We can also guess that speech input is going to be harder to process than textual input. This is because speech adds another layer to the process. Speech is really spoken text, so turning the speech into text and then understanding the meaning of the textual messages is a two-stage process or at least a more difficult job than beginning with the text in the first place.

Speech processing shows us yet another example of the inverse perception problem. If we wish to generate speech from text, then by encoding a set of phonemes and generating suitable sounds through various hardware means, it is relatively easy to produce spoken messages. This can now be done on many microcomputer systems. But the inverse problem of receiving, understanding and turning the messages into textual form is much harder. We have an analogous situation with vision where we can easily generate a two-dimensional picture on a screen from computer data about a three-dimensional object, but we find the reverse process of interpreting three-dimensional data from two-dimensional images very hard. Thus, the philosophical or strategic problems of perception which arise out of the inverse speech problem are very similar in character to those in the inverse vision problem.

In Figure 7.2 some application areas for the four different possible combinations of input and output are listed in an approximate order of difficulty. In the first case, *processing textual input into speech*, we can see application areas here, both real and potential, in all sorts of talking systems. These will comment and communicate to the user, while the user replies through coded signals or other non-verbal input media. The talking computer terminal would be an example here. There is a great deal of interest in such developments particularly for use in intelligent computer-aided instruction which is a very active research field at the moment.

The second possibility of *processing text into text* is seen in the classic example of foreign language translation systems. In the early days of AI, this was one of the first research areas

Operation Mode	Example Applications
1) Text to Speech	Talking manuals, Tutoring systems, Alarm and warning systems.
2) Text to Text	Language translation systems, Database query systems, Intelligent editors.
3) Speech to Text	Voice driven typewriter (talkwriter), Speech controlled appliances.
4) Speech to Speech	Dialogue systems, Telephone information systems, Intelligent advisors and assistants.

Figure 7.2 Language systems

to be tackled and there were many spectacular failures due to the simplistic approach of mapping words across from one language dictionary into another. Hence, phrases like 'hydraulic ram' became translated into 'male water sheep'. However, most of the early mistakes and difficulties have now been overcome, and less ambitious, but more useful, translation systems are now available. Another example of text-to-text processing is seen in database query systems where natural language front-ends provide user-friendly access facilities.

The third area, concerning the *conversion of speech into text*, is best illustrated by the example of the voice-driven typewriter or 'talkwriter'. This is a project which has long been under development at IBM and there is no doubt that there is a tremendous potential market in worldwide sales of 'talkwriters' when they are eventually perfected. IBM have demonstrated a talkwriter which is able to transcribe complete business letters using a vocabulary of 5000 words. The Japanese are also very active in this area and have produced impressive systems. It seems certain that talkwriters will eventually become widely available and their impact could be very significant. They will replace most typewriters and, in combination with word-processing software, should enable very rapid document generation and manipulation. It is quite likely that one's own personal talkwriter will become the main input device for all sorts of natural-language processing. In effect, the keyboard, as it is known today, will become redundant. However, IBM alone have been working on this for over fifteen years and it is clear that there are many problems. We should not expect the results to be available as a commercial system for some time.

The fourth possibility is *speech-to-speech systems*. These systems will be able to provide a complete spoken dialogue. This will be useful in all manner of conversational roles. Computer 'assistants' will talk to people about problems, and advice and information will become readily available. An example might be a computer switchboard operator that (or who!) will answer telephone queries. These systems will be created for areas where human operators are currently employed purely for message passing and related types of communication. The conversational role of such systems makes heavy demands because in order to maintain a conversation of any complexity about a particular subject requires both a knowledge of that subject and also some understanding of the intentions and purposes of the other speaker. Thus, some form of model of the speaker must be developed. This will require knowledge to be maintained about the wishes, beliefs and goals of the participants in the conversation. Clearly, the task then becomes much more than a conversion or interpretation process.

All of these areas present quite severe research problems but they are gradually being developed. At the present time there are many research groups working on these topics. These include groups at Edinburgh and Cambridge in the United Kingdom, SRI, MIT, CMU and IBM in the United States, and several major European and Japanese projects. It is quite clear that, when speech systems become very fluent and natural-language processing systems can be offered with power and flexibility, there will be tremendous commercial markets to be gained. Of course, they will not appear overnight as the problems are very difficult and we should only expect incremental improvements. When remembering the difficulty of the inverse problem we realize that the latest successes and achievements manage to reach a viable solution by using highly restricted domains and introducing constraints of various sorts. For example, if the domain of discourse is contained and restricted to a particular topic, then knowledge about the context of a message or conversation is much easier to provide and can be made very effective in guiding the interpretation process. Thus, we see emerging commercial systems which can only talk about particular applications or problem areas, we experience some limiting input restrictions and we see various stylistic controls and constraints on users actions and inputs. This is in accordance with our familiar trade-off: if the problem or system is constrained, we gain some measure of success in a limited world, but every time we wish to lift a constraint to increase performance we need a yet more powerful theory to provide us with the tools to handle the new problem areas.

7.2 Speech processing

We sometimes think of speech as an inferior ability compared with vision. We know that vision is two-dimensional and involves colour, depth, motion and other parameters, whereas speech appears to be a serial activity — one word follows the next — and the complexity at any one moment only concerns the understanding of a given word or phrase. However, this apparent simplicity hides many complications. Although sound is transmitted by pressure waves in air and therefore appears to be a one-dimensional variable, the pressure varies dramatically over time. Figure 7.3(A) shows a plot of a sound pressure wave as sensed at a microphone. The actual trace of the microphone voltage (representing pressure) is really not much use as it is the pitch or frequency of the various notes that conveys the information. In Figure 7.3(B), a frequency plot is shown which indicates the amounts of different frequencies contained within the signal. This breakdown of a signal into its different components is known as a spectral analysis. However, such frequency plots only show a particular instant in time, and during a speech act the frequencies will change constantly giving a dynamic frequency pattern. Consequently, we need to re-introduce the first dimension: time. One of the best ways of examining speech in terms of magnitude,

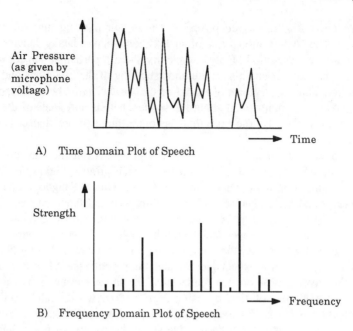

Figure 7.3 Sound measurements

frequency and time is by creating a sound spectrogram. Figure 7.4 shows a typical spectrogram. This is a kind of three-dimensional diagram, produced by a machine, known as a 'spectrograph', which repeatedly scans a piece of sound recording and measures the energy in different frequency bands. The frequencies are plotted on the vertical axis, with time along the horizontal axis, and the darkness of the trace indicates the amount of energy at a particular frequency/time point. In this way, spectrograms usefully capture the patterns of change in sounds or speech vocalizations. Notice the complexity of the spectrogram, which only covers a few seconds of speech. Although rising or falling tone patterns often appear as a characteristic of speech, the structure of speech utterances is nearly always highly variable and very complex. The spectrograph is a very useful tool for the analysis of syllables and phonetic structures.

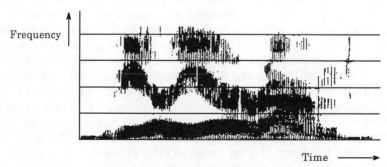

Figure 7.4 A sound spectrogram

In order to capture and store speech of reasonable quality we would have to take digital samples of the signal around 10 000 times per second. This means that speech passages lasting only tens of seconds would very rapidly fill up megabytes of memory. Thus, as for vision, we have a selection and analysis problem. What is required is the identification and extraction of only the vital measurements and parameters that we can use to analyse the sound pattern and detect the word structures that we are looking for. However, this is not a simple analysis technique — it is not possible to design an algorithm which will extract a small set of parameters that are sufficient to interpret any given English word. Different renditions of the same message by different speakers can produce almost totally dissimilar sound structures. Conversely, two almost identical sound patterns can have quite different origins *and* meanings. In order to understand a message we may need to analyse it at many different levels. We will also need to know the full context of the message, i.e. the surrounding structures that have been detected and the relationships between them. The acoustic or sound properties of a message represent only one dimension of a multi-layered problem.

Figure 7.5 shows the different areas that we will be interested in studying when trying to analyse human speech content. Ranging from the most primitive levels based on acoustic variables we go through phonetics and lexical analysis (word structures) to syntax (word sequences), semantics (word meanings) and finally on to the purpose and the intention of the message. The first three areas are specific topics that have to be studied for speech processing. This is the reason why speech processing is potentially more difficult and complex than textual language processing which is 'only'(!) concerned with lexical analysis onwards. Clearly, there are many different aspects which must be considered here and many of these areas overlap and interact with one another. Perhaps the most obvious approach would be to work up from the acoustic level, through the phonetic layer, through syntax and semantics, until we eventually reach a complete understanding of a message. However, this type of sequential processing through layered hierarchies is only sufficient if the message is clearly defined and we have no difficulty in understanding its nature and context. In

Acoustics	*- properties of sound and its transmission*
Phonetics	*- the study of families of primitive sound fragments*
Morphemics	*- the study of word sound formations and groupings*
Lexical Analysis	*- the detection of individual words*
Syntax	*- the analysis of word sequences with respect to a grammar*
Semantics	*- the analysis of word meaning*
Pragmatics	*- the analysis of the purpose or intention of the message*

Figure 7.5 Levels of language analysis

PERFORMANCE GOALS	RESULTS FROM HARPY	RESULTS FROM HEARSAY II
Accept connected speech	Yes	Yes
From many speakers	3 male, 2 female	1 male
Quiet room environment	Computer term. rm.	Computer term rm.
Allow slight system tuning per speaker	20 training utterences per speaker	20 training utterences per speaker
Allow slight training of speakers	none	none
Vocabulary of 1000 words	1011	1011
Highly artificial syntax	Context free semantic grammar	Context free semantic grammar
Highly constrained task domain	Document retrieval requests	Document retrieval requests
Less than 10% semantic error	5%	9%
Sentence error	9%	19%
Number of test sentences	184	22
Processing limited to a few times real time on a 100 million instructions per second computer	28 million instructions per second of speech	85 million instructions per second of speech

Figure 7.6 Results from ARPA speech understanding project

practice, we soon discover that this rather mechanical bottom–up analysis turns out to be far too simplistic.

A different method is to independently interpret each of the layers using separate processing modules. Each process then works on the relevant parts of the input data and tries to propose its own hypothesis for the input message. When there is any significant measure of agreement then the message can be considered understood. One of the best known systems which uses this style of independent processing is the Hearsay-II Speech Understanding System from Carnegie-Mellon University.

During the early 1970s the American Defense Department sponsored four large speech-understanding programmes through its Advanced Research Projects Agency (ARPA). The projects ran in competition and were evaluated after a five-year time scale. Both Hearsay-II and another system from Carnegie-Mellon, called Harpy, generally satisfied the requirements specification and were considered successful. Figure 7.6 shows some results from these systems.

The achievement of these results was significant because they demonstrated the feasibility of computer understanding of connected speech. Prior to these projects the speech problem was considered too difficult for existing technology. While there are some shortcomings in the results, the requirements specification was very strict and the two listed projects achieve the overall goals of recognizing connected speech with low error rates. The error rates are of two kinds: semantic errors — where the output (a document retrieval request) did not match the intention of the speech input, and sentence errors — where the generated document-retrieval request is not word-for-word perfect with the input sentence but essentially generates the correct semantic request. A comparative analysis with the other two projects in the ARPA programme suggests that the strong constraints on vocabulary size, syntax and task domain were major factors in helping Harpy and Hearsay-II to achieve success.

Apart from its pioneering speech processing results Hearsay-II has become a classic study of the *blackboard style of problem-solving architecture*. We recall from Section 6.5 and the example in Figure 6.22 that blackboard systems are useful for combining heterogeneous kinds of knowledge in a cooperative regime where different processing modules (knowledge sources) manipulate data in a global data base (the blackboard). Figure 7.7 illustrates the blackboard structure and the knowledge sources employed in Hearsay-II. The different levels in the blackboard deal with quite different forms of knowledge about speech structure. This is a characteristic of blackboard systems. In addition to knowledge sources which process information *between* the levels there are also knowledge sources which process data *within* levels. The knowledge sources are logically independent and can process data at any time provided they are triggered by events on the blackboard and have sufficient input data to provide useful processing. In order to control the system there is a knowledge scheduler which activates the different knowledge sources and selects them for execution according to the state of the blackboard. Such schedulers are necessary in serial computer systems in order to share out the processing resources. However, it is possible to build a blackboard architecture in which parallel knowledge sources operate entirely independently. Such systems are now being built for experiments, often using different computers or processors for the various knowledge sources.

Although the different knowledge levels handle heterogeneous types of information, blackboard knowledge is usually related in an hierarchical organization. For example, knowledge of syllables can be used to produce results for input to a word-analysis process and word information can be used in processes that generate word-sequence structures. In this way, information is passed up and down the hierarchy of levels in order to help solve problems that cannot be solved within a given level. The system builds up partial solutions all over the blackboard and these are gradually completed by the action of the knowledge sources filling in the details in a cooperative style of interaction. Thus, the blackboard system is both an integrated approach to problem solving and is cumulative or incremental as it continues to build solutions on the blackboard as and when data are available. This style of operation is called *opportunistic scheduling*; both top–down and bottom–up processing can occur on the blackboard hierarchy at any stage. The scheduling process implements the idea of least commitment in order to keep all solutions open until enough data are available. The final interpretation of a speech utterance may depend upon many hypotheses and these can all be maintained on the blackboard and the choice between them delayed until sufficient information has been generated to aid the solution.

To summarize, blackboard systems are useful for problems which have poorly-defined goals and multiple representation spaces that can be brought to bear on the problem. They are valuable when an incremental, opportunistic resolution of multiple hypotheses is

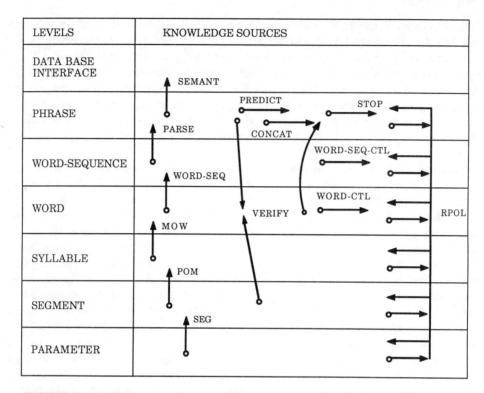

KNOWLEDGE SOURCES
- SEG: Digitizes the signal, measures parameters, and produces a labeled segmentation.
- POM: Creates syllable-class hypotheses from segments.
- MOW: Creates word hypothesis from syllable classes.
- WORD-CTL: Controls the number of word hypotheses that MOW creates.
- WORD-SEQ: Creates word-sequence hypotheses that represent potential phrases from word hypothesis and weak grammatical knowledge.
- WORD-SEQ-CTL: Controls the number of hypotheses that WORD-SEQ creates.
- PARSE: Attempts to parse a word sequence and, if successful, creates a phrase hypothesis from it.
- PREDICT: Predicts all possible words that might syntactically precede or follow a given phrase.
- VERIFY: Rates the consistency between segment hypotheses and a contiguous word-phrase pair.
- CONCAT: Create a phrase hypothesis from a verified contiguous word-phrase pair.
- RPOL: Rates the credibility of each new or modified hypothesis, using information placed on the hypothesis by other KSs.
- STOP: Decides to halt processing (detects a complete sentence with a sufficiently high rating, or notes the system has exhausted its available resources) and selects the best phrase hypothesis or set of complementary phrase hypotheses as the output.
- SEMANT: Generates an unambiguous interpretation for the information-retrieval system which the user has queried.

Figure 7.7 Hearsay blackboard system

appropriate using different sources of knowledge drawn from related but separate aspects of the problem. Blackboard systems reduce uncertainty by integrating knowledge.

Since the seminal work of Hearsay-II, the following ten years has seen many developments and experiments with blackboard systems. There are at least twenty significant applications, extensions, and refinements of the basic idea in the literature and there are also several attempts at generalizing the blackboard architecture to form a general methodology of integrated problem solving. One of the most promising and appropriate applications appears to be the sensor fusion problem in which signals arriving from diverse sources are to be integrated and coordinated to produce meaningful messages about the semantics of the sensed readings. Many applications are concerned with military problems of target identification and vehicle recognition but there are also applications in robotics which are well worth pursuing. The problems of sensor signal integration was mentioned in Section 2.3 and is one of the most pressing and important problem in sensory interpretation. Blackboard systems would appear to be an ideal vehicle for experiments in this area.

7.3 Text and language analysis

One of the big success stories in natural-language analysis is the emergence of commercial machine translation systems. There are now several companies which will offer foreign language translation packages. The success of these systems is really based on the relaxation of the requirement for total automation. These systems will translate between a specified language pair, say German to English. A typical system uses something like 20 000 rules for each language pair — far less knowledge than that of a good quality human translator. The results are not completely translated but contain marked sections (around 15% of the text) which require post-editing by an operator in order to iron out various nuances, ambiguities and complexities that could not be efficiently handled by the package. These systems are used to help improve the quality and throughput of translators, without taking over their role. This is an example of the combination of human and machine giving better results than either of them on their own. It is interesting to note that most of the new firms in this area have been set up specifically to manufacture and market machine translation systems. This indicates their confidence in the development of the technology and its likely acceptance in the marketplace.

Another area of success in language processing has been the development of intelligent front-end programs for databases. These front-ends allow queries to be directed at a database in natural language. Such systems are now commercially available and are becoming popular. IBM, for example, uses the large and expensive Intellect system which has been increasing its market share by a rapid growth in the number of installations.

Figure 7.8 shows two typical database queries in a conventional query language format together with their English equivalent. It is clear that a naive user would not be able to design formal queries like this without a fair amount of prior training. The English language versions, however, are very easy to comprehend. It is the job of these intelligent front-ends to convert such English statements into formal queries for the database. Notice that there are two levels of knowledge involved here. The system must not only know enough about English to determine what the user wants but it must also have considerable knowledge about the query language and the available database facilities in order to formulate a suitable query that will satisfy the users' needs. This type of product is a very useful facility as large databases have many fields with complicated indexing methods, many data record structures and quite complex organizations. A simple English query input would be very valuable and allow many more users access to databases than through formal query methods. Although such systems were developed originally for mainframes, like the Intellect system, there are now many microcomputer versions available and it is quite likely that they will become an integrated part of most future database systems.

```
                    DATABASE STRUCTURE

                   ENTITIES AND ATTRIBUTES

    MACHINE ( machine-name, location)
    COMPONENT ( component-no, component-name, drawing-ref, machine-name)
    MATERIAL ( component-no, type, cost)
```

QUERY 1   "Find the names of all components with at least two
               materials"

               SELECT component-name
               FROM COMPONENT
               where component-no =
                     SELECT component-no
                     FROM MATERIAL
                     GROUP BY component-no
                     HAVING COUNT (type) >= 2

QUERY 2   "Find the names and drawing references of all
               components being machined at machine 'NC7' in
               material 'Brass G7'"

               SELECT component-name, drawing-ref
               FROM COMPONENT
               WHERE machine-name = 'NC7' and component-no =
                     SELECT component-no
                     FROM MATERIAL
                     WHERE type = 'Brass G7'

Figure 7.8   Database interrogation

7.4 Robotics and factory systems

In Chapter 1 we suggested a scenario in which a conversation took place between a human operator and an industrial manufacturing cell. We have now reviewed some of the technology that will be required in such conversational systems and it is the purpose of this section to consider the effectiveness and appropriateness of such modes of communication in future manufacturing systems. The problem of communicating with computers is a major topic and the research area known as *human computer interaction* (HCI) is a multi-disciplinary area involving psychology, ergonomics (human factors), mathematics, physics and computer science. While such topics cannot be dealt with here we can review the developments in speech and language processing and make observations on the suitability

```
            INPUT

     Binary acknowledgement of signal
     Selection from one of n options
     Enter scalar quantities
     Enter geometric/vector quantities
     Enter formal text ( e.g. task program )
     Enter semi-formal text ( e.g. specification )
     Enter natural language


            OUTPUT

     Binary status indicators
     One from n selections
     Display scalar quantities
     Display geometric structure/quantities
     Give general explanations
     Display informative text
     Issue priority warnings
     Broadcast alarms
     Directed messages
```

Figure 7.9   Categories of message structure

of these for the requirements of HCI communications within the manufacturing arena.

Figure 7.9 illustrates the different kinds of message structure that can pass between human and machine. These range from rigidly-controlled signals in a prescribed format to unconstrained natural-language inputs. We notice that as input and output constraints are lifted so the complexity of the message can increase. In many manufacturing situations it is sufficient to instruct and control a process by selecting from a limited variety of possible options. The output responses from machines and processes also often range over a small set of pre-coded messages. Most of the commonly used methods of control and communication with factory machines fall into this category, which can be classified as limited option selection and response. We see concrete examples of this kind of message structure in the use of switches, buttons, lights and menu displays. Other more complex message types are indicated in Figure 7.9.

Figure 7.10 illustrates the different modes by which these messages may be transmitted. Here again, the opportunities offered by some communication modes are very limited whereas others allow free-flowing and complex input or output sequences. We notice that the display screen is an almost universal output device as it can simulate many other forms of output, including dials, gauges and other conventional visual indicators. Displays can also show photo-images and these may be produced directly from scanning sensors or could be processed and enhanced in some way. For example, when an inspection process has detected a faulty component, the location and nature of the fault could be highlighted by an enhanced image illumination level. Considering most manufacturing activities, it seems that

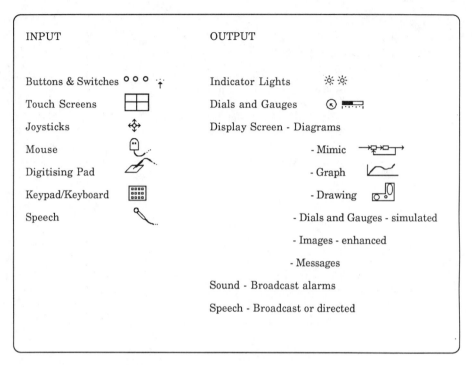

Figure 7.10   Modes of communication

either menu-type selections are to be made or directional or geometric information is to be processed. There seems to be very little need for free-flowing natural language or speech as used in normal human communications. If we look back at the original scenario in Chapter 1 we see that nearly all the messages that passed from the operator to the machine could have been accomplished through simple menu selections, and numeric displays would be sufficient for the machine's messages in the other direction. The inspection stage used vision data but this can be dealt with separately, using its own display screen, as an independent sensor. So there really does not *have* to be any form of speech communication path for the tasks illustrated. However, this is not to say that conversational systems are undesirable, just that they should only be introduced when there are clear benefits. Even in cases where speech facilities seem desirable, there will be many psychological barriers and ergonomic problems which will have to be overcome before these systems are accepted in everyday life. There are many examples of advanced systems failing to be accepted by their operators — not through any lack of technology, but because they have not matched the needs or expectations of the operator's cognitive style.

The task of programming say a washing machine or a video recorder through spoken commands is not quite as easy as it sounds. A considerable degree of precision is necessary regarding operations, times, and sequences, and these require a degree of accuracy that the machine can understand without the ambiguity that is in so much natural language. Hence, it is likely that the designers will introduce a semi-formal spoken language that constrains the message variability while focusing on the tasks to be defined. The many different speakers who issue the same commands, but by using entirely different sounds and word sequences, will still provide many exciting problems to be solved! In fact, the multiple-speaker problem may well be the biggest difficulty in the whole system. Of course, the present method of programming through a bewildering array of buttons and knobs is also a semi-formal approach to the task and is also not ideal for human interaction. Nevertheless, switches and buttons are attractive to designers because they are *precise* and have no scope for vagueness or ambiguity. It is interesting to notice the introduction of bar-code readers that can be used to program video recorders by scanning the details from television timetables and magazines. This method might become more convenient for the user than programming by speech. One area where speech seems to offer strong benefits is in the spoken acknowledgement of commands. If our washing machine or video recorder could verbally summarize the program it is about to perform, we might find this more convenient than examining various dials and displays.

Let us now consider first of all speech and then natural language in the manufacturing context. Speech processing will be useful in certain situations. For example, spoken alarms are very effective at attracting attention, and broadcast messages, directed at individuals, are able to override visual stimuli. Such spoken warnings, generated by speech synthesizers, are now being used on aeroplane flight decks to alert crews to critical situations. Another useful application is where an operator requires both hands for a particular task and can speak to a piece of equipment to give it additional commands. A crane is an example of this where simple commands — up, down, left, right — can be used to guide the crane while the operator is physically engaged on other tasks. Commercial systems such as the Verbex 1800 speech recognizer are now available and these can recognize words and messages of a form suitable for many such control tasks. Some of these systems are also speaker-independent, in that they do not need to be trained on different speakers. However, the success of many such systems is based on a rather small vocabulary, typically only around 100 words. As the vocabulary size increases so it is more likely that substitution errors will be made in which the system confuses one word with another. It is to be noted that in several tests on existing systems, done by independent parties, quite serious confusions were produced through substitution errors. For example, a common substitution for 'stop' is 'start', and a substitution for 'no' is 'go'! It is important to view these commercial systems as word recognizers rather than as speech understanders; the difference is the same as that between pattern recognition and scene understanding, as was stressed in our discussions on vision. Their preference for isolated words rather than normal connected speech, the need for training sessions, the limits on multiple speakers, and their small vocabulary, all show that these commercial systems are only partial solutions to the complex problem of speech understanding. When these points are considered, together with the problems of noise in the environment, and various other extraneous difficulties that may occur with speech, it is questionable whether the freedom and the scope for complexity and confusion that speech input offers is really beneficial for much of the manufacturing domain.

Regarding natural-language processing, there are also considerable difficulties here because the intentions and desires of the speaker are not always clearly expressed in the message. False starts are very common in human communication. The messages are quite often vague, and arbitrary pieces of information arrive in a fairly random or jumbled sequence. This is not a conducive framework for receiving precise instructions to control manufacturing systems. In short, errors and ambiguities occur at all levels of natural language and speech communications and these have to be successfully handled and resolved by any intelligent system that is interpreting such messages.

If we consider the knowledge that is required in such a system, there are at least three levels. The first level involves the knowledge of dialogue and the rules of discourse that are involved in the current conversation. The second level concerns world knowledge about the process or events that are being discussed in the dialogue, in our case, this would be details about the manufacturing process. Finally, the highest level involves knowledge and beliefs about the person who is being spoken to in order that reasons for the purpose and aims of the statements can be understood. A model of the other person's goals and beliefs is necessary because the motivation of the participants in conversations has to be known, or at least estimated, in order to reason about the context of messages and to make sense of any missing information.

There are only forty phonemes involved in English speech and perhaps ten thousand syllables but a typical speaker will know at least two hundred thousand words. When these are presented in combinations of unlimited variety it is simply not sufficient just to look for word matches in an attempt to try to map incoming utterances into meaningful knowledge-based statements. With these points in mind it seems unlikely that natural-language input is a desirable feature for manufacturing systems. The aim of manufacturing is to automate as much of the production process as possible. In other areas, total automation is frequently avoided because a passive role for operators can be dangerous. In situations such as aeroplane piloting, a totally passive role leads to inattentiveness, lack of motivation and, ultimately, negligent performance. However, in a production environment we are trying to manage our information as automatically as possible and we do not wish to have humans involved in the control loop. We would like to have much high-quality information available for decision making in other areas, particularly for plant management, but we do not wish to slow down the production process by having decisions made at a low level where the processes will be held up by human communication. One positive possibility for natural-language output is in computer-generated reports and summaries. Written explanations of system behaviour and readable reports on technical issues will be valuable user-friendly services that aid training, management and maintenance.

Due to the precise nature of most engineering data, the inherent flexibility and variability of natural language is often a disadvantage rather than a benefit. While natural-language communication is unparalleled as a flexible means of message passing, the need for precision in engineering really emphasizes the difficulties that any language recognition system faces. We notice also that as more and more of the manufacturing process becomes computer-based, then more and more of the data will have to be machine readable. In other words, we desire the inputs and outputs of all our machines to be compatible and we wish to develop a completely integrated manufacturing environment. To this end, local area network protocols such as MAP and TOP are being developed and used for inter-machine communication. It is absolutely certain that future factories will have powerful and extensive information flow around the manufacturing plant with a great deal of management data being available. The complete factory will operate on the basis of computer-generated data and there will be very heavy traffic between all the parts of the manufacturing system. We would expect that the factory floor with its various automated manufacturing processes will require very little in the way of natural human communication, and displays and sophisticated menu-driven processes will be sufficient in order to satisfy all monitoring and supervision requirements. When speech and natural language is viewed in this context, it is clear that it will have a rather minor role on the factory floor. On the other hand, these facilities will be more valuable in areas where the messages are less precise and less critical. Such applications are found where humans and computers cooperate in problem-solving tasks. These include all 'computer-aided' areas, including computer-aided design, computer-aided manufacturing and all kinds of decision support systems. Most of these applications will be concerned with management and early engineering design decisions. Consequently, we expect most speech and natural-language facilities to be found at a fairly high level in the manufacturing system, and they will mostly be used to enhance the user-friendliness of management and operational decision-making activities.

7.5 Summary

1 Speech and natural-language processing are, like vision, very hard problem areas. One reason is that we are without a good understanding of human perception.
2 Significant advances have been made in speech and natural-language technology. Language-translation systems, intelligent database front-ends and speech synthesis and recognition systems are now commercial products.
3 The advent of the talkwriter will eventually replace the keyboard as the main computer input and interaction device.
4 Conversational systems need to have three layers of knowledge: the rules of dialogue, subject knowledge and an understanding of participants' intentions and desires.
5 Hearsay-II demonstrated an incremental, integrated, opportunistic approach to speech understanding and established the blackboard model as a major AI methodology.
6 Natural-language inputs will be inefficient for factory floor operations which require precise messages.
7 Speech outputs will be important for high priority warnings.
8 Computer-generated natural language text will be useful for reports and explanation facilities.
9 Speech inputs will be valuable in office environments, especially for interactive systems such as CAD and decision support systems.

7.6 Further reading material

For a collection of classic natural-language papers see *Readings in Natural Language Processing*, edited by B. J. Grosz, K. S. Jones and B. L. Webber (Morgan Kaufmann, 1987).

Computer Speech Processing, edited by F. Fallside and W. A. Woods (Prentice-Hall, 1985) gives a broad introduction to work in both speech analysis and synthesis.

For a description of the Hearsay blackboard model, details of the other systems in the ARPA project, and the final results, see D. H. Klatt, 'Review of the ARPA Speech Understanding Project', *Journal of the Acoustical Society of America* (1977), volume **62**, number 6, pp. 1345–1366. Erman *et al.* give full details of Hearsay-II in the reference cited in Section 6.7.

The original paper on the sound spectrograph, in the *Journal of the Acoustical Society of America* (1946), volume **18**, number 1, pp. 19–49, by Koenig, Dunn and Lacy, gives insight into the nature and physics of speech and sound.

Chapter 8

Emulating the expert

An expert is a man who has made all the mistakes, which can be made, in a very narrow field.

Niels Bohr

Gummidge's Law: 'The amount of expertise varies in inverse proportion to the number of statements understood by the general public.'

A. Bloch

In recent years there has been much excitement about expert systems and their applications in industry and commerce. In this chapter we will look at the principles and performance of expert systems and see how they fit in with the requirements for automated manufacturing.

8.1 The basic expert

The concept of an 'expert system' is a relatively recent idea, although its roots have a long history in AI. An expert system is a piece of software which is able to perform in a rather narrow but detailed area of specialist knowledge. The essential aim of an expert system is to act as a consultant and provide useful advice and expertise in particular specialist fields. Two commonly-cited fields are: medicine (where an expert system might help to diagnose the nature of a disease through an interactive session with a doctor), and geology (where data from geological surveys are used by mineral or oil-prospecting expert systems to deduce the location of valuable deposits). In fact, most of the exciting examples cited in the literature are research systems and are still under development and refinement. In the marketplace, many systems have been sold, but most of these are used for evaluation and experimentation; only a small, but rapidly-growing, proportion are used on a regular basis as an established part of company business. Recently, there has been an explosion in the number of commercial products that are offered by software houses and other vendors. These usually take the form of expert system 'shells': systems with in-built representation and inference machinery but empty of application knowledge. Shells are programmed for specific applications by feeding in rules or other formulations of the available expert knowledge.

Expert systems are nearly always directed at areas that lack a fully-developed model or theory, i.e. areas that are only partially understood. If this were not the case more formal analytical methods could be used. So they are aimed at 'fuzzy' areas where everyday rules of thumb and other heuristics have evolved. This does not mean to denigrate such topics. Indeed, human performance can be very impressive in areas rich in experience but weak in scientific principles. Perhaps the main contribution of expert systems has been to open up this very area, i.e. to automate human reasoning in applications that have no formal theories. This is why certain parts of medicine are good examples, because *complete* models of patients and diseases are not available — nor are ever likely to be!

The results of an expert system are usually accepted as judgements rather than exact factual results. This is in accord with the analogy made with human experts. Humans are good at weighing evidence to reach quite precise conclusions, often without giving any exact numerical quantities. The aim is to make such experts' knowledge more readily available throughout the industries in which they work. It is a surprising fact that the most skilled top-class experts are actually quite few in number. Because of this, they have to travel extensively and use all kinds of media in order for their skills to reach the problem areas. Consequently, the idea of a software package containing expert knowledge and being readily available to many people is very attractive. In theory, this should allow better medical treatment, better diagnosis of failures, better quality products, and, in general, an opportunity for enhancing human decision-making skills. Such systems will not have 'off' days, will not default to rival companies and will always be on duty. Notice that we do not suppose that humans, expert or otherwise, will necessarily be replaced by these systems. Because of the inexact judgemental nature of the output, the role for most expert systems will be to act as an assistant in improving the quality of an area of work, rather than as an absolute authority.

Just because commercial expert systems are readily available, as are expert system tools to develop one's own expert system, does not mean that the expected benefits will necessarily be easy to realize. Indeed, there is a great deal of exaggeration about the potential benefits of expert systems and some wild claims have been made. We will review some of the techniques here, examine their shortcomings and consider ways in which they could be improved and be made more suitable for industrial automation problems.

Figure 8.1 lists some of the main features that might be found in a typical expert system.

Rule Based (often production rules + probability weightings)

Large quantity of Domain Specific Knowledge

Flexibility in Knowledge Base Structure

Multi-faceted Knowledge Base

Traceable line of Reasoning

Good Explanation Powers

Ability to Acquire, Refine & Organize Knowledge

Figure 8.1   Expert system characteristics

First, they are often implemented by using rule-based designs. This usually implies a production system architecture. The rules are frequently enhanced with some form of probability measures that represent confidence levels for the various relations and facts (see Section 5.4). These values are used in inference procedures to enable the evidence from several rules to be combined, so giving more reliable results and generating recommendations about their certainty or doubt. In particular, the statistical method of Bayesian inference is a popular technique for combining evidence. Secondly, a great deal of domain knowledge has to be coded into the system. Vast amounts of knowledge are involved in any detailed expert domain and expert systems cannot avoid or ignore this. Thirdly, the knowledge base must be flexible because facts might be removed, new facts may be introduced and amendments or updates to the knowledge may take place at any time. Also the knowledge base could be multi-faceted, that is, it may contain similar knowledge in several different forms. For example, a system might have information about a particular disease at various levels of description: the biological level, the symptom level, the treatment level, the relationship with other connected diseases, and so on. Another feature is that the reasoning which is used in expert systems to produce their results must follow a recognized approach that can be understood by people. Thus, it is not only the results that are required, but also a record of the development of those results by a well-defined line of reasoning. The next point bears on this as the system should be able to explain how it reaches its results, and the explanation facilities should provide clean, easy-to-understand reports for human consumption. This relates to the human interface aspect of computers and, clearly, good user-friendly facilities are required in expert systems. Finally, any knowledge-based system should be able to acquire, refine and organize its knowledge on a continuous basis, and expert systems will be expected to accommodate new data learned during usage sessions. This means they should automatically maintain their knowledge base as effectively and as efficiently as possible.

One of the big problems with expert system work is locating the knowledge in the first place. The knowledge-acquisition problem, sometimes called the 'knowledge-acquisition bottleneck', was outlined in Chapter 5. The standard technique is to find one or more really good experts in the field and have long discussions about their information, their reasoning techniques, and their opinions. It is very desirable for the expert to see the results of an early expert system trial. The expert's comments will invariably bring about corrections, perhaps involving quite drastic changes. The system thus evolves through a long iterative process of design, evaluation and modification. The reason why it is done this way is that there is no known formal technique for knowledge acquisition. In fact, many experts are quite unable to explain many of their lines of reasoning including how and why they came to particular views. Most human experts learn very slowly and incrementally and, by the time they are expert, they are not sure of the procedures that they used to become experts. The process of encoding knowledge into expert systems is often called 'knowledge engineering'. This is an unfortunate usage; it would be better to define knowledge engineering as 'applied AI' and encompass all aspects of application-based knowledge processing. All professional designers and builders of knowledge-based systems could then truly be called 'knowledge engineers', not just the encoding technicians of expert systems. After all, knowledge is a central concern of the whole of AI, not just expert systems.

There are at least three different modes in which an expert system can be used. First, the expert system can solve problems as an expert. This is the most common mode, in which the user is a client and employs the system for advice and consultation. Secondly, the user could be seen as a tutor, in which case the system receives knowledge in order to patch up holes in its knowledge base, improve its performance, and generally take advice. Finally, the user could be seen as a rather naive pupil. In this role the system displays its knowledge in a suitable form and the user learns through structured instruction sessions. This mode is also useful for examining and cataloguing the system's knowledge.

One of the most well-known industrial users and developers of expert systems is DEC (Digital Equipment Corporation). DEC has been using an expert system as a component of their manufacturing operations since 1980 and has many others under development. Their philosophy for introducing expert systems into new areas hinges on four basic guidelines. When looking at a new task area, they ask the questions:

1. Are people good at it? — i.e. implicit knowledge already exists.
2. Is the knowledge largely empirical? — i.e. no theory has been formulated.
3. Can heuristic rules be applied? — i.e. useful relationships have been observed in the data.
4. Is less than 100% success acceptable?

In DEC's view, if these criteria hold, then it is probably worthwhile developing a prototype system and investigating an expert system for serious use. These guidelines hint at the crucial nature of expert system domains — they are concerned with areas where *implicit* knowledge already exists but there is not enough *explicit* data for manipulation in formal, theoretical solutions. Implicit knowledge is information that people have developed, often over several years, is usually not recorded anywhere, but is intrinsically involved in their activities, actions and decisions. By its very nature implicit knowledge is not accessible. This is one of the main difficulties that expert systems attempt to deal with. Because implicit knowledge is generated by people and is embedded in their procedures and conventions, it often has the characteristics of being ill-specified, contradictory, incomplete and inexact. By dealing with this kind of problem, expert systems create formulations of explicit knowledge that were previously only known implicitly. The exercise of generating explicit knowledge is becoming seen as the most valuable part of any expert system project, especially in the long term. The spin-off of creating explicit data for what was previously known only implicitly turns out to be far more important than the existence of a new consulting expert system. This tends to be because the many applications and new potential uses for the data only emerge after it is seen and appreciated.

Unfortunately, implicit data are not the kind of material that science likes to work with. Consequently, although there exist a large number of case studies with various examples of implicit knowledge being used, we do not yet have a science or an underlying theory for building expert systems for given applications. There is a great deal of unknown territory but researchers are trying to formulate sound methodologies and map out their understanding. Out of the collective experience of many different expert system experiments we see the emergence of some general principles for designing expert systems. These can be summarized as follows:

1. It is generally a good principle to separate the domain knowledge base from the reasoning machinery.
2. It is a good idea to use uniform representations, even for multi-faceted layers if possible.

3   It is useful to encourage and exploit redundancy wherever possible.
4   Reasoning techniques should be no more complex than the minimum required by the problem.

8.2 A few difficulties and the need for research

At present, systems have been produced which will solve basic domain problems, explain their reasoning in a rather stylized manner and are sometimes able to perform limited learning. However, we are a long way from the ideal expert system. When we look at the operation and performance of a typical package we often notice serious shortcomings. It is worth reflecting on these in order to determine future progress.

During operation we find that the explanations for decisions are inflexible and lack insight into the domain. Explanations are usually generated by reading rules in reverse. So the reason why a particular disease was suggested will be explained by giving details of the symptoms associated with that disease. This is really only a trace of the execution of the expert system and falls far short of the detailed causal analysis expected of a consultant.

Another problem is that expert systems do not degrade gracefully when the quality of their knowledge becomes low. Usually when they run out of a line of reasoning they simply stop. They do not give messages such as 'I feel rather unsure of my reasoning in this area, I advise you refer to another source of expertise, such as … '. This type of response would be more useful than the usual sudden collapse or, even worse, completely wrong advice.

There are other serious difficulties in getting expert systems to restructure their knowledge, to be able to discover exceptions to rules, and to determine the relevance of different rules in different contexts: all features of human expert reasoning. Any learning facilities are usually fairly crude and consist of rules being replaced with more precise or more specialized rules.

There is also a problem in being sure that the knowledge acquisition process has been well done and has obtained the most suitable data for the job. Partly due to this, it is very difficult to validate systems formally in order to guarantee their behaviour. Formal validation procedures are desirable for most software packages and expert systems are no exception. We are still a long way from a set of scientific principles that will allow us to prove how a given expert system will work or to verify a design or performance specification. In addition, the requirement for flexibility means that knowledge-base structures will change from day to day and so it will be quite difficult to compare systems on the basis of their performance in order to justify different design variations.

Some of the reasons for these problems that are experienced when building and designing expert systems are due to the factors listed in Figure 8.2. Taking these in turn:

1   Firstly, most problem domains involve large numbers of facts; thousands of rules are involved in any reasonable expert system. In a recent system, a model of cardiac disorders of the heart included over 140 000 rules. This is also true in engineering where a typical component may require many hundreds of lines of data just to describe the geometric information in an engineering drawing. There will also be much associated data concerning component features, their relationships, tolerance data, and the materials and their properties. All these data have to be coded into suitable representations and the sheer volume may inhibit structural improvements or experiments with alternative reasoning methods.

Emulating the expert

Large quantities of information - thousands of facts

Noisy data - inaccurate, incomplete and unreliable information

Time varying problems - dynamic structure or parameters

Tentative reasoning - variable beliefs and unstable assumptions

---

Figure 8.2 Problems with expert systems

2 Another problem is that the data are usually noisy. This means that there are missing facts, some facts are in conflict with others, and some facts are purely extraneous. This leads to conflicts and unreliable data. This is why probability reasoning methods are such a major feature in many expert systems. Simple arithmetic averaging will fail as a method of weighing evidence in the presence of noise, whereas Bayesian and other statistical methods can effectively reduce the effects of uncertainty. However, statistics cannot replace large areas of missing knowledge and we still require deductive and other inference methods for dealing with systematic patterns of missing or conflicting knowledge.

3 Another important topic, which most expert systems entirely ignore, is the problem of time-varying problem features. In dynamic situations, such as plant control and medicine, the set of facts which is true at any one time may vary considerably. And so plants move into different modes of operation, diseases move into different stages and the parameters which were previously reliable suddenly become irrelevant or unreliable. This is another major research area in expert systems and an important aspect of AI. Some of the research on game playing has produced methods for changing sets of parameters when a game enters a new phase.

4 Finally, the limited style of reasoning in most systems is a major shortcoming. Two different but related kinds of reasoning illustrate the problem. Assumption-based reasoning is a technique whereby certain facts are assumed to be true without the support of absolutely concrete evidence. This implies that, at a later stage, the facts may change as we discover more details about them. Such ideas have led to the development of truth maintenance and belief maintenance systems which have special in-built mechanisms to handle updates on the knowledge base. These are becoming more sophisticated and truth maintenance systems may be used in many future expert systems. The other kind of reasoning concerns the tentative aspect of many decisions made by experts, in other words, the first thoughts which occur and are then later discarded. This closely relates to assumption-based reasoning techniques. Many AI techniques, such as constraint propagation methods, can involve tentative arguments. Both tentative reasoning and assumption-based reasoning require the maintenance and manipulation of different contexts or viewpoints. For example, a number of alternative tentative scenarios may be generated, each being incompatible with the others, but we may wish to maintain all of the scenarios unless they get eliminated by being inconsistent with the current data. Such

```
Basic Issues
    - The Knowledge Acquisition Problem
    - Powerful Explanation Facilities
    - Validation Procedures
Major Enhancements
    - Strategies & Plans
    - Temporal Reasoning
    - Spatial Reasoning
    - Combining Related Knowledge
    - Conflict Management
    - Learning from Experience
    - Using Analogies
    - Subjective Reasoning
    - Focusing Mechanisms
```

Figure 8.3   Research areas for improved expert systems

context-handling mechanisms are now being built into new AI tool kits, and these are being used in experiments that will influence the next generations of expert systems.

All in all, these shortcomings mean that there is much research to be done if expert systems are going to deliver the promise which many of their suppliers and manufacturers have claimed. Figure 8.3 indicates some of the relevant research activities that will help with these problems. These topics are all under active investigation and will be important in improving the performance of existing expert systems. For example, most systems do not have an internal strategy for what they are trying to achieve, and important features such as temporal reasoning and spatial reasoning are quite rudimentary at the moment. Conflict management is the problem where several different rules or methods have produced different answers and are to be reconciled. Subjective reasoning and focusing mechanisms are concerned with the ways in which people concentrate on selected areas of a problem, and choose particular lines of argument. All these topics are likely to see new developments in the future and will then lead to much better products.

This section has shown that many complex and interrelated issues hide behind the label 'expert systems'. A great deal of active research is currently investigating these areas.

Diagnostic Tasks	Design Tasks
Analysis of system state from observed variables	Synthesis of system state from action sequences
A limited range of states are caused by many possible events	A limited range of actions can produce many possible states
Determine what action sequence caused the current state	Determine what actions should be used to reach a desired state
Backward reasoning required from current state to past states	Forward reasoning required from current state to future states

Figure 8.4 Expert system roles, analysis versus synthesis

8.3 Expert systems in industrial automation

Expert systems seem to fall into two broad families: *diagnostic systems* and *design systems*. Diagnosis closely relates to monitoring and sensory interpretation while design subsumes topics such as prediction and planning. The primary functions of diagnosis and design correspond to analysis and synthesis respectively. Figure 8.4 shows some characteristics of design contrasted with those of diagnosis, and indicates the different styles of problem, input data and reasoning involved.

In diagnosis an expert system is used to suggest reasons for a failure or malfunction of some kind. The diagnosis role can be extremely useful in all sorts of factory automation where complex machining systems, robot systems, and other manufacturing machinery are involved in high-quality production. As soon as a failure is sensed then an expert diagnosis system could evaluate the situation and save much damage or down-time. The expert system could be linked to the target machine over a network or might even be in-built as part of the machine's control. Of course, the response from the expert system and its method of operation would have to be tailored to the requirements of the shop floor, the factory control computer and any supervisory or management system that is to be used. This mode of operation would be an on-line mode. More common is the off-line mode of diagnosis, in which batches of data from previous tests and measurements are processed by an expert system to try to determine the cause of a failure. This mode can involve consultation sessions with experienced personnel.

All kinds of diagnosis have been subjected to the expert system approach; in manufacturing all areas that affect productivity and quality control, such as the analysis of manufacturing parameters, failed products, materials or machines, are being thoroughly investigated.

Regarding design expert systems, these will become extremely useful in computer-aided design (CAD) where the designer requires help and information to support the feasibility of a particular design. Although geometric data may be the main language of CAD, designers also need to analyse material properties, estimate the costs involved, and decide on the actual production operations. A suitable expert system closely interlocked into a CAD system could provide this information interactively, so that as the designs evolve, so also do the feasibility and production requirements emerge as well. Such an expert system could give advice on designs difficult to manufacture and make recommendations for improving structural, functional or production aspects of the design.

It seems that a merger of both human and machine would be the most beneficial arrangement here. For design, we do not propose a completely automatic system which designs the final products without any intervention by people. The computer can efficiently generate and suggest alternatives, but the evaluation of the most subtle parameters, which are very difficult to define in machine terms, can be best left to humans. These include aspects such as aesthetics, variables relating to cost and production methods, and factors concerning the history of the product line and its competitors. On the other hand, expert systems for robot and machine control should be as automatic as possible because consulting sessions will be inappropriate during production activities. Not only would this waste valuable manufacturing time but the nature of the tasks are machine-based, rather than human-centred. Thus, there are two kinds of system for manufacturing: fully automatic knowledge-based systems, e.g. for monitoring and control, and interactive expert systems, e.g. designers' assistants.

In addition to product design, expert design systems could be found in areas such as computer-assisted assembly where robots are used to assemble parts in different sequences. Here, advice on difficult assembly sequences could be used to assist the flow of work through assembly cells.

The power of such design tools will depend on the size and quality of the available knowledge and databases and the range of reasoning methods built into the expert system. Rather than trying to build one all-embracing expert system, a community of cooperating expert systems, that are individually optimized for different tasks but share common design data and communicate results, seems a more promising and productive approach. In the same way that a distributed data-processing system is able to draw on data from different points within a business network, so it will be possible to gain expertise from different aspects of the manufacturing process and bring these to bear at any given problem point.

Although diagnosis and design are two major kinds of expert system, other, closely-related, types of expert system are also of interest in the manufacturing cycle. Figures 8.5–8.7 show how different styles of expert system can be used at various stages in product manufacturing systems. Figure 8.5 schematically shows the stages involved in organizing and operating a manufacturing plant of the type discussed before. The eight small rounded boxes indicate different opportunities for expert systems to provide support for the main manufacturing process.

A product begins life at the design stage when product requirement documents initiate the complex creative activities that result in a series of product specifications. The specifications define both the product in detail and some of the processing necessary for its manufacture. Most products consist of a series of components that can be manufactured separately and then combined in an assembly stage. The strong opportunities for expert systems to support existing CAD were mentioned above.

From the various design documents and specifications we next have to organize a series

Emulating the expert

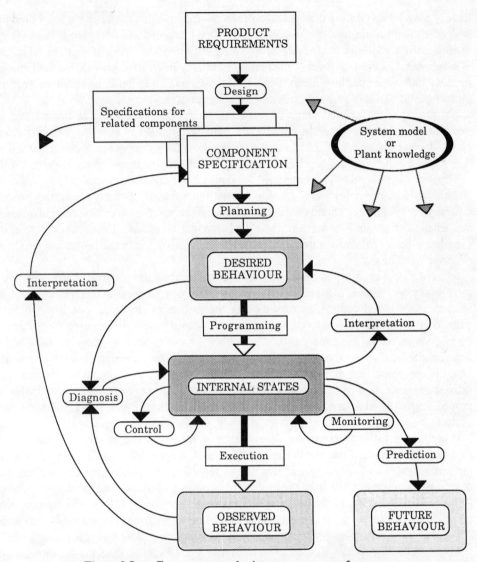

Figure 8.5 Expert system roles in component manufacture

of manufacturing activities that, when carried out by the factory, will result in production of the designed components. This is a planning task and is one of many different levels of planning found in manufacturing. Our immediate concern is the planning task that determines the sequence of manufacturing operations that must be performed to produce the components. This is called 'process planning' and is carried out by engineers who specify the correct series of processes, machines and tools that will create the component parts. Process planning works from the viewpoint of the developing parts and considers their processing needs and manufacturing sequences. Other forms of planning deal with ordering raw materials, specifying production quantities, and scheduling resources and deliveries. The production department deals with such tasks and will try to balance the loading of the factory plant by scheduling the operation of the different machines and processes. Often this is done through a master schedule and the factory resources and component routings are organized by a centralized hierarchical 'material requirements planning' system. This kind of planning takes the viewpoint of the factory machines and tries to provide for their smooth running and even loading. Many expert systems have been built to assist or perform planning in all kinds of planning task and for all areas of production control.

The output of the process planning stage is a description of a series of actions that the manufacturing plant is to perform on the raw material. In other words, this is the statement of the desired behaviour of the plant machinery. In order to make the equipment perform this desired behaviour we need to write a series of precise instructions to program the equipment. These are the 'part programs' for numerically-controlled metal cutting machines and the manipulation programs for robot handling systems. The programming process represents a kind of translation of the desired behaviour into the necessary internal states so that when the machines enter these states the behaviour is realized. Hence, execution of the machinery corresponds to transitions through the specified internal states and produces observable behaviour.

In most cases, the desired behaviour will be identical to the observed behaviour and a satisfactory operation is achieved. However, there are many opportunities for error and we wish to monitor and control the execution of the system as closely as possible. This is a situation for execution monitoring expert systems that dynamically observe the steps and event changes as the machinery proceeds through its cycle. Such systems are designed to detect variables that could be moving into critical states. Control expert systems use similar kinds of variables (but, this time, including control signals) to maintain the system on a satisfactory execution path. Both monitoring and control are very much concerned with the immediate low-level activities of the plant and are applications for on-line automatic expert systems.

When some failure occurs we detect it by seeing that the observed behaviour is significantly different from the desired behaviour. A diagnosis system is able to measure these differences and may generate hypotheses regarding the abnormal internal states that caused the plant to deviate from the desired behaviour. Expert systems for diagnosis of all forms of manufacturing machinery are being extensively researched. This is a commercially-significant area of work as accurate and rapid diagnosis gives considerable economic benefits in maintaining machinery and keeping manufacturing plant fully operational.

There are other expert systems shown in Figure 8.5. Prediction systems are those which are able to speculate about the future behaviour of a system or plant. Prediction systems are valuable because they may indicate future areas of operation where system variables could become critical. They may also be used to answer questions about different modes of operation. Prediction systems will usually entail a simulation facility so that instead of experimenting directly on the plant and then observing whether it is operating efficiently or effectively, we can use a prediction expert system to simulate the behaviour without expensive trials or damage to the equipment.

The remaining expert system shown in Figure 8.5 is an interpretation system and this attempts to deduce, from observed behaviour, the goals or intentions of the manufacturing plant. Two versions are shown. One is working backwards from observed behaviour to the desired goals, in other words, a kind of reverse planning system. The other one is shown

Emulating the expert

interpreting internal states in terms of the intended behaviour. This is a sort of reverse programming process in which the actions performed by the system are interpreted to give the instruction sequences. Interpretation is a very general term and many expert systems could be seen to be solving interpretation problems. Diagnosis, for example, can be viewed as a specific kind of interpretation problem.

We notice that several of these expert systems will require detailed knowledge about the manufacturing processes. Some of them will need a specific model of the manufacturing plant, as in the case of prediction, while diagnosis, for example, requires a detailed understanding of the way in which different behaviours can be generated by the internal states of the machinery. A model is a coherent body of information that captures the essential aspects of a concept or entity and provides enough detail to support reasoning

Function	Features
DESIGN — Configure options under constraints	Many sources of constraints; Global & local optimisation - by redesign; Sub-design factoring gives structure; Decision justification required; Spatial reasoning involved
PLANNING — Recommend actions to achieve goals	Event planning - choose activity sequences; Location planning - choose spatial positions; Motion planning - choose collision-free paths; Must consider uncertainties; Difficult to plan for sensors
MONITORING — Detection of abnormal conditions	Sensor coordination problem; Alarm management involves partial diagnosis; Significance of parameters varies with time; Execution monitoring akin to opportunistic planning
CONTROL — Regulate system behaviour	Performance optimisation; Sensor coordination problem; Trend analysis; Feedback through learning
DIAGNOSIS — Create failure analysis from observations	Requires model of system; Requires access to critical variables; Single, multiple & intermittent faults; Must allow for errors in observations; Should degrade gracefully
PREDICTION — Infer consequences of current states	Requires model of system; Simulates system operation; Involves temporal/sequential reasoning
INTERPRETATION — Infer goals from actions	Noisy input - weighted evidence methods; Partial data - extrapolation methods; Conflicting data - belief based reasoning; Unreliable data - assumption reasoning

Figure 8.6 Functions and features of manufacturing expert systems

ACTIVITY	INPUTS	OUTPUTS
DESIGN	Product requirements, product constraints, knowledge of manufacturing capabilities	Product/process specifications
PLANNING	Product/process goals & specs., process knowledge, production knowledge	Action or event sequences
MONITORING	State variables, critical regions, system knowledge	Performance indicators
CONTROL	State variables, control parameters, system knowledge	Control variables
DIAGNOSIS	Observed behaviour, desired behaviour, system knowledge	Localised fault hypotheses
PREDICTION	State variables, system knowledge	Future performance
INTERPRETATION (1)	Observed behaviour	Behavioural goals
INTERPRETATION (2)	State variables	Model of system

Figure 8.7 Expert system functions

from first principles. Expert systems that directly interact with the system, such as control and monitoring, can manage with knowledge bases that do not have a complete model and may be based on heuristics and other approximate information. It is not quite so vital for these expert systems to have a complete system model. However, for higher level activities such as design and diagnosis, the lack of a model or equivalent detailed plant knowledge is likely to be responsible for very poor performance.

Figure 8.6 illustrates some of the features of these different classes of expert system and notes their various functions, requirements and important attributes. Figure 8.7 lists the inputs and outputs of the expert systems. Input/output relationships are another way of looking at systems and help us to understand their purpose and aims.

We notice that yet other types of system could also be proposed. For example, following the diagnosis of malfunctions, a debugging expert system could recommend different remedies that are suitable for the diagnosed malfunctions. From these recommendations, a

Emulating the expert

Figure 8.8 Information flow in manufacturing

repair expert system could then generate an execution plan in order to administer a remedy which has been selected by the debugging expert system.

These are just a few examples closely associated with actual production but there are many more that indirectly support production. If we consider the whole process of manufacturing, not just the engineering aspects, we find many situations where expert systems could make a contribution. Figure 8.8 shows the many business, engineering and management activities that surround the main production loop. Any one of these areas may offer information and knowledge problems which might well be effectively handled by the introduction of a suitable expert system. We have only examined the areas concerned with production and robotics but prototype systems have been built and studied in all these other different application areas.

A current problem with many operations involving small batch manufacturing and flexible production plant is the dynamic nature of much of the production data. Either the data are not completely known at the time planning is to commence or the information changes during planning and execution. This means that flexible planning systems are needed which can rapidly change and adjust as more information becomes known. This style of planning is known as 'opportunistic planning' and is closely related to execution monitoring. This kind of monitoring or planning is very much the form of control expertise that is needed in many manufacturing operations, including flexible robot assembly.

In recent years some factories have moved from a rigid hierarchy of top-down planning of materials and processes (the 'material requirements planning' systems, mentioned above) into a more demand-driven system, sometimes known as 'just-in-time' scheduling, whereby items are manufactured as and when needed. Material requirements planning is a top-down, vertical hierarchy, where parts are 'pushed' through the factory, while just-in-time and related KANBAN methods are distributed, bottom-up planners that 'pull' the parts through their processing. This style of organization offers economic benefits as it reduces the amount of inventory and tries to optimize the work in progress. However, the control is more distributed and less easy to manage from a single point of authority. In this framework we see opportunistic planning systems as being highly relevant. Small changes due to errors and machine failures in localized cells will introduce noise into any hierarchical planning system and the idea of a single masterplan can then become unworkable. In fact, many factories use a number of different redundant paths for parts manufacture simply because various historical accidents have introduced additional different part routings which have then become established. It is often easier to continue with such inefficiencies than to stop everything and try to sort the whole thing out from a fresh start.

In modern environments, execution monitoring, powerful diagnosis systems for error and fault detection, and rapid flexible opportunistic planning systems will have a major role to play. These systems will also make more explicit information available and, as mentioned earlier, this will be one of the main benefits. Expertise and information which was previously locked away in engineering practices or in subconscious staff behaviour will become an accessible commodity. This also allows better decisions to be made and it may well be that the major future value of all of these systems is not so much in their immediate and local effects as in the understanding and explicit information which they generate for enhanced management understanding, better plant operation and improved product design.

From this brief survey of the manufacturing cycle we see many different layers of expertise which are open to enhancement by expert systems and other forms of knowledge-based system. In the factories of the future we imagine many such systems linked through communication paths so that they may cooperate and reinforce each other in order to produce high-quality products which are better designed and rapidly manufactured by flexible factory systems.

8.4 The generate-and-test approach

We now describe a major technique that offers a powerful basis for several of the above expert systems. This technique, known as 'generate-and-test', is very useful in both design and diagnosis problem solving. The basic idea sounds almost trivial: think of possible configurations for the given problem and examine them to see if they could provide a solution. This is equivalent to a search through the problem state space with a test for goal states. However, the power of the method lies in the effective application of constraints to reduce search effort. Generate-and-test is most useful where *all* solutions are to be found, for example, in diagnosis problems, where all possible scenarios for the failure or abnormal situation are to be discovered. The classical example here is in medical diagnosis, in which all possible diseases need to be considered for a particular relevant set of symptoms.

In order for generate-and-test to work, the state space must be factorable into smaller spaces so that the generator can prune out whole *classes* of solutions and reduce the candidates to manageable numbers. Any available constraints, from the problem specification or application data, are used to eliminate classes of solutions. In this way, the generation process creates all the *plausible* candidates for a given situation. It is best to

Emulating the expert

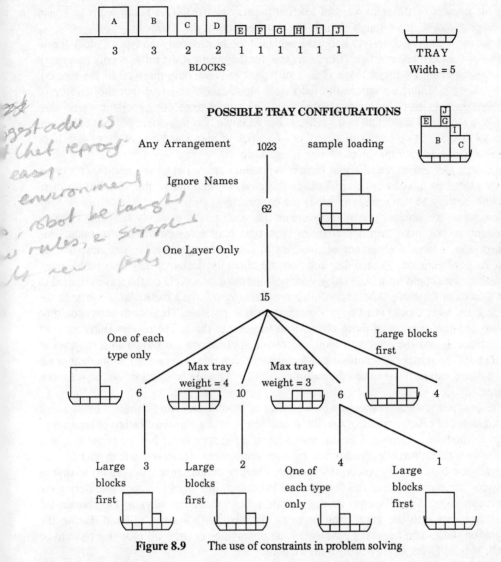

Figure 8.9 The use of constraints in problem solving

consider an example and once again we will use our robot world with another case from the heat treatment plant. Figure 8.9 shows 10 different blocks that can be loaded onto a tray. There are three types of block, six small ones, two medium-sized and two large ones. Assume the cell designer/programmer wishes to instruct the robot to load the tray and wants to find a suitable combination of blocks. The designer asks the computer to produce all feasible configurations of tray loading. (Notice that this design problem could easily be recast as a diagnostic problem: for example, assume the robot has found a tray in the oven left by a maintenance crew, and the actual configuration of blocks on the tray is to be diagnosed.)

First, if we do not know of any constraints at all then any logical combination of blocks could be on the tray. Assuming that we identify each block separately with its own name, then the number of different ways of loading the tray is 1023. This is the worst case and corresponds to a search through all possible arrangements of the 10 blocks.

However, we may not *need* to know the names of the blocks; only the type of block has to be known for the robot grasping operation (and, for diagnosis, if the robot is only to remove the tray contents to a reject bin it is also sufficient to know only the type of the blocks). Using this constraint, we ignore the individual block labels and find that the number of solutions reduces to 62 different possibilities. The output from our generator would thus suggest a stream of candidate tray loadings. For example, the tray could contain either one block of size 3, or two blocks of size 3, or one block of size 3 and one block of size 2, or two blocks of size 2, or one block of size 1, and so on.

But more constraints are likely in most real problem areas. Let us suppose that no block can be placed on top of any other block. In this case, we find that of the 62 solutions there will now be only 15 that remain to satisfy the constraints.

Now let us add another constraint. Assume that each feeder can only feed one block at once, and so not more than one block of each type may occur on the tray at once. This immediately reduces the number of solutions to six. Our expert system can now display these six configurations as candidate solutions for either the design or diagnosis problem.

Further cases could have been suggested; perhaps the total weight on the tray is limited to 3 or 4 units, or there might be a scheduling requirement to force a particular block on to the tray, e.g. the large blocks must be processed as soon as possible. These constraints could be applied in place of some of those above or in addition to them. The relationships between the constraints and the effects of their successive applications are shown in a diagram in Figure 8.9. The number of solutions can sometimes reach one or even zero. In the latter case the designer can remove selected constraints until a suitable solution or set of solutions emerge.

This example shows the way in which a range of possibilities can be pruned by virtue of constraints. The often dramatic reduction in problem size is a valuable feature of constraint-based methods of reasoning. This can only happen if the state space has the property that it can be factorized into regions defined by such constraints. However, where this is true, generate-and-test is a very powerful technique. The key points for success are (a) to find as many constraints as possible and (b) to utilize the constraints as early as possible during the generation stage. The efficiency of the method relies on finding ways to incorporate the constraint data into the generation process. If constraints are not utilized *during* the generation stage, then too many candidate solutions will be created, all of which have to be tested. This will lose all the power of the method.

8.5 Towards deeper levels of understanding

There has been some disappointment with the results produced by many systems. After purchasing an expert system shell or development kit, the user loads in all the knowledge that is available, and then sits back for a high-level consultation session with the new instant expert. The true story is really never like this. Most expert systems produce a very naive response, and the ones which are truly successful, i.e. are heavily used, have many thousands of rules and are continuously being developed, even while they are in use.

The problem seems to lie in the nature of the knowledge and the way in which it is stored.

Many expert systems use rules, heuristics, and probability relationships that might be labelled 'surface knowledge' or 'shallow knowledge'. This type of knowledge gives information about *observed properties* of applications. There is often no indication of how those application systems actually operate; the data and rules only indicate external relationships and external associations. Clearly, with that type of information there will be serious shortcomings when deeper understanding is required. For example, in a diagnostic task, a new mode of failure may have arisen which has never been seen before and in this case it would be no good looking for a suitable rule, or even combination of rules, that would explain the new situation. When a new disease suddenly appears (e.g. AIDS) it's no use trying to treat it as a combination of existing diseases; we must go back to basics and identify the bacteria or virus and investigate its effects and features. What is required here is a return to first principles so that the reasoning follows from a basic understanding of the mechanisms *inside* the application. Research is now being directed at techniques known as *deep knowledge* or *deep reasoning*, in contrast with existing *shallow knowledge expert systems*. This rather grandiose sounding title is really an umbrella term for studies that cover many different aspects of advanced knowledge representation and reasoning. It concerns fundamental concepts such as causality, intention and basic physical principles in order to reason about the way devices operate and systems perform. From this viewpoint we see that heuristics are really short cuts; they represent the input/output relationships of quite complicated systems and avoid the deeper reasoning necessary to deduce the machinery behind those input/output relationships. In other words, they give the final conclusions of other, more fundamental, reasoning processes.

This is not to say that heuristic rules are unnecessary or unimportant. Such pre-compiled knowledge is very useful and can be very efficient in cases where they apply. They also save a great deal of unnecessary re-working of previous reasoning. However, unless *all* relevant rules have been captured there will be situations where this type of system will be unable to produce a sensible response. Whenever human experts find their heuristics are becoming inadequate, especially when new or unfamiliar cases arise, they seem able to switch to a more analytical style of reasoning which draws on first principles. This is one of the characteristics of an expert — the ability to creatively analyse a new situation. It is the aim of deep knowledge research to try to build systems with similar abilities.

Deep knowledge methods depend upon powerful models that contain details of the basic principles of organization and operation of the application. They involve some form of notation that can express the fundamental nature of the domain. In other words, the system is able to appreciate the basic principles by which the data are derived. In this way, not only can the system reason from symptoms to causes but it is possible to pre ´t future behaviours that those causes could be responsible for. In addition, the effects of structural or other major changes to the problem domain can be investigated by re-working the expert analysis from first principles.

Let us consider the example shown in Figure 8.10. Here we have a small part of an electronic circuit. In case (A) we have some heuristic rules derived from observations of what happens when certain voltages are measured at certain places in the circuit. This will be adequate for representing how the circuit behaves during those modes of operation. However, a set of measurements might be experienced that are wildly different and therefore will not match any of the rules. This corresponds to a new mode of behaviour that has not been covered by the heuristics. The expert system will fail to give a diagnosis here

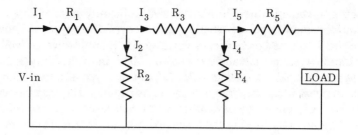

A   SURFACE RULES

(voltage V-in 10) & (current I_1 50)   →   assert (voltage LOAD 1.25)
(voltage V-in 10) & (current I_1 44.5) →   write(LOAD disconnected)
(voltage V-in 10) & (current I_1 52.4) →   write(warning LOAD short circuit)

B   DEEPER CIRCUIT KNOWLEDGE

(current ?x ?y) & (resistance ?x ?z)   →   assert(voltage ?x (divide ?y ?z))
(joins ?x ?y ?z) & (current ?x ?m) & (current ?y ?n)
                                           →   assert(current ?z (difference ?m ?n))
(loop ?x ?y ?z) & (voltage ?x ?p) & (voltage ?y ?q)
                                           →   assert(voltage ?z (difference ?p ?q))

Figure 8.10   Electronic circuit analysis

or, even worse, will give a completely wrong evaluation. By contrast, this problem could be dealt with by a rule base of the type shown in (B), where the underlying physical properties of such circuits are encoded. The rules describe Ohm's Law and the current and voltage addition laws, that is, they encode some basic circuit theory. Notice that the topology of the circuit being analysed will be held as connection assertions in the system's working memory, for example:

(joins I1 I2 I3)
(joins I3 I4 I5)
(loop V-in V1 V2)
(loop V2 V3 V4)
etc.

Using these rules the system can work out the necessary voltages and currents throughout the circuit in order to analyse any unusual values or departures from normal behaviour. If the circuit topology changes, i.e. a structural change in the system occurs, then the second rule base can still be used to good effect. In the first case we would have to add new rules for each case of changed circuit topology. This would cause the system to require constant modification and growth to deal with new situations (as has been experienced in many

Emulating the expert

expert systems). The disadvantage with the deeper approach is that much more work is needed during reasoning, as each case is treated from first principles. Notice that the two formalisms both use a very similar rule structure and, of course, it is not the rules themselves that matter, it is the contents of the rules which determine whether a system is a shallow or deep-knowledge-based system. If it is a shallow system, it will only have input/output associations, whereas if it is deeper it will model the structure of the system and represent internal cause-and-effect relations. Both of the two styles may be implemented using any of the representation methods in Chapter 5.

A set of characteristics can be outlined to capture the idea of deep knowledge systems:

1. Deep knowledge entails models reflecting the structure of the domain. As indicated above, design and diagnosis problems will benefit from models of the implementation level of the application.
2. Structure should be distinguished from function. This characteristic is phrased differently by different researchers, but the general idea is that the representation of the parts of a system should not assume the purposes for which they are to be used. This 'no function in structure' principle recognizes that the function or operation of a system must be separated from its physical structure. In diagnosis problems erroneous events are likely to cause structural changes leading to components performing novel functions, and yet we still wish to deduce how they may behave. If operational data are tied into structure we will have to anticipate every type of error behaviour for every type of structural fault. This is very inefficient and is a major cause of poor performance in shallow systems.
3. Causal relationships. The representation of relationships between different components or processes in the deep knowledge model reflects the underlying reasons why the shallow heuristic rules usually employed by experts are applicable in certain cases (and why they are not applicable in others). Because of the fundamental nature of this type of reasoning, this is sometimes referred to as 'reasoning from first principles', or even 'common sense reasoning'.
4. Deep knowledge models are different in kind from numeric models; each has its advantages and disadvantages. The reasoning done by deep knowledge systems tends to be symbolic rather than numeric in nature. Such systems attempt to manipulate higher-level representations in the same way as humans seem to do, sacrificing some of the accuracy of a numeric model. They try to produce explanations and justifications at a more appropriate level for human consumption than is the case with mathematical models of a domain. The problem with numerical methods is that although they are very accurate such precision is often unnecessary for general common sense diagnosis. The exact values are not important, it is really the mode of behaviour or the mode of failure which matters. With qualitative reasoning, explanations can be based in terms of general domain concepts, without computing large sets of numerical quantities for each different structural configuration that might be proposed.

Because there are many different ways of looking at a problem area, it is often very fruitful to entertain many different representations of what is essentially the same application. Consequently, there may be several different layers in a deep knowledge base. Figure 8.11 shows an electronic circuit. In addition to the normal circuit diagram, several

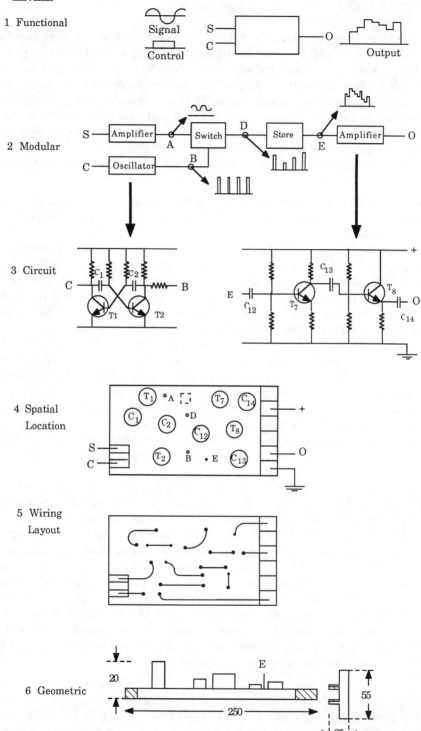

Figure 8.11 Layers of electronic knowledge

other representations are also shown. A schematic diagram indicates the overall function of the system, while a block diagram identifies the logical modules. A spatial layout diagram shows how the parts are laid out on a printed circuit board and a wiring diagram gives the physical positions of the wired connections. Finally, a geometric diagram gives precise dimensions of the printed circuit board and components. All these representations refer to different views of the same circuit but they are all very useful in diagnostic reasoning, troubleshooting, design and analysis. By moving from one level to another the focus and perspective of a problem can be altered to deal with other aspects.

Consider a troubleshooting scenario based on this diagram. If the circuit was not performing correctly we may first look at the block diagram to consider which part of the circuit was causing the trouble. If there was no output at all, we may look along the block diagram, to consider which stage was faulty. This is helped by the provision of test points in the circuit. In order to read the voltages at the various test points we would have to locate them on the circuit board, and so we will need the level 4 representation. We might reach test point E and find that there is a signal, but at point O, there is no output. This observation allows us to reason that the output amplifier is the site of the fault. Next, we can look at the circuit diagram to find the section that implements the amplifier and deduce what could be wrong. At this level we will need to know some electronic circuit theory, and, as a result of an analysis, we might suspect that transistor T_8 was the most likely to be damaged. We could then go back to the schematic diagram of the board, locate that transistor and measure its operating parameters. We notice that one of the voltages is too high and also, while we were running the test, we noticed that T_8 gets very hot. We now have to look for reasons why this voltage could be too high. Further examination of the wiring diagram indicates that the gaps between that part of the circuit are very small and on closer inspection we can see that there are faults in the track on the printed circuit board.

So we have reasoned through from the function of the circuit down to its possible behaviours, located the faulty components by noticing their behaviour being in error, and then found reasons for the error in the physical level of the implementation. We might then suggest changes or recommendations to the designer which involve geometric considerations and changes to the circuit board. On the other hand, we might consider that the electronics could be improved instead and change some of the component values or some of the actual circuit design itself.

In such general diagnostic reasoning, we tend to flit backwards and forwards between different kinds of knowledge. However, this is not an *ad hoc* process; we are following a line of argument (or perhaps several arguments at once) and building up an understanding by integrating pieces of information from several related sources. Thus, the different layers are used in an integrated manner to support common sense reasoning processes.

8.6 New techniques for mechanical systems

A simple mechanical device serves as an example to illustrate how these ideas can be applied to the mechanical domain. Figure 8.12 shows a small piece of mechanism consisting of five parts. When force is applied at L the trigger T pivots at 'p' about the frame F and applies pressure to the block B. As B is fixed to the rod R a force is experienced and the rod moves against the spring S into the positive region of N. Notice that qualitative values, rather than numeric values, are being used; the variables have values

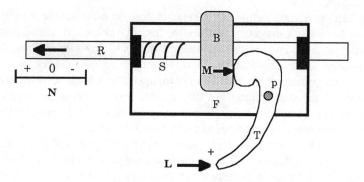

Figure 8.12 A mechanical example

+, 0 or –, representing their broad qualitative nature instead of precisely measured quantities. In this example, we are using qualitative values for force and position: +L means some non-trivial level of force exists in the direction towards the right and +N means the rod's position is near the left. Hence, the normal operation of our mechanism, 'pressure at L leads to rod being in position N', can be expressed as:

+L → +N

The action of the spring gives other behaviours:

0L → –N, and –L → –N

The diagnostic problem for this device can best be appreciated by analogy with the task faced by human maintenance engineers. The observable variables are L and N; these are the diagnosis inputs. We assume that the internal states of the device are not available, indeed, descriptions of these form the outputs of a diagnosis session. The diagnostic task is to infer reasons for the behaviour of the observable variables in terms of internal states that could support that behaviour. A maintenance engineer would be trained on the device and would have knowledge in the form of an internal model of the device and its different modes of behaviour. During diagnosis the engineer would formulate questions about the observable variables and suggest states for hidden components that might be causing the faulty behaviour. The failure region might be narrowed down to a limited area and then various test procedures could be suggested to gain further relevant data. For example, a test might involve removing a cover over the spring to confirm a proposed error state — broken spring.

A diagnostic expert system might be constructed for this kind of device by extracting heuristic rules from maintenance engineers and then coding these into a rule-based framework. Thus, two sample rules for different error cases are:

IF –L AND +N THEN spring S is broken
IF +L AND –N THEN block B is disconnected from rod R

However, this approach leads to the sort of expert system, as described earlier, that depends upon shallow knowledge. Many researchers are now trying other methods for building expert systems which have a better understanding of mechanical and electronic systems.

The author's research group is working on the design of systems that have no such pre-compiled rules but instead attempt diagnosis by simulating the operation of the device. These systems infer faults by identifying components that have to be altered in the simulation in order to produce results that match the observed faulty behaviour. Instead of using the repair engineer's heuristics that are gained through long experience, we wish to simulate the mental predictive reasoning that is provoked by novel failures (and is also often seen in less experienced engineers).

To build such a deep expert system for a given problem area, we first gather data about the structure and function of the given circuit or mechanism, most of which would be readily available from engineering and design data. This knowledge is then used to build a model containing the causal and physical relationships involved in the working of the device. Then, by propagating qualitative values through the structural model, a set of experimental reasoning processes are able to generate possible device behaviours. This kind of 'qualitative simulation' of the real system has been called 'envisioning'. The generation of behavioural scenarios from structural data is a key feature of first principles reasoning systems. In diagnosis, envisionments can be used to match anticipated error behaviour with observed error behaviour in order to deduce the likely internal situation. However, it is not necessary to generate all possible error behaviours as there will be sufficient available knowledge about the design and structure of the device to formulate constraints that restrict the scope of the envisionments. The diagnostic system will assume that most of the normal structure is intact until proved otherwise. By using the natural device hierarchy that is inherent in most engineering machinery (and, hopefully, has been retained in our representation) it will be possible to propagate normal (and sometimes faulty) variables inwards from the environment to successive levels of device detail. This should localize the section of the structure where errors are being generated. The next task is to match the localized symptoms on to patterns of behaviour that the system recognizes as characteristic functions. Such 'operational schemas' will associate fragments of behaviour with the device functions that cause them. In this way, we can propose device changes that might be responsible for producing the error result. In some cases there will be several possible outcomes and hence multiple diagnoses will be reported, but if the level of detail is sufficient there will be enough constraints for only very few schemas to match: sometimes only one. The argument here is that mechanical devices are designed in a systematic manner to produce highly organized and constrained behaviour and diagnosis should utilise that design and organizational knowledge to help understand deviations from the designed behaviour.

In order to model mechanical devices in this way, the knowledge base will require considerable depth, probably with multiple levels and different styles of representation. As a minimum, this will involve structural knowledge, in terms of spatial relations and interactions, functional knowledge, in terms of input/output and cause/effect relations, and also some design data regarding intended behaviour. In addition, some sequencing or elementary time reasoning will be needed, as will knowledge about physical properties and physical events such as the relative strengths of materials, the effects of friction and the nature of component interactions.

With the system described, we aim to show how an expert system can produce behaviour

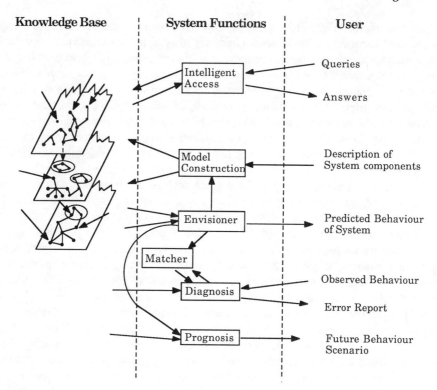

Figure 8.13 A multi-layered expert system

that approaches 'understanding'. The system should be able to perform detailed simulations of possible behaviours using structural and operational hypotheses. By reasoning about the properties and features of different components it will deduce candidate scenarios for faulty behaviour. When observed events appear incompatible with normal operation, basic physical knowledge can be used to suggest structural changes that could account for the deviation. In this way, our system should handle broken springs, bending triggers, fractured rods or distorting frames, in addition to less dramatic failures, such as the block becoming detached from the rod (the case of +L −N), or seizure of the pivot. We suggest that this approach is a start on the path towards expert systems that truly understand as much about a given mechanism as an expert diagnostician.

Figure 8.13 shows a possible design of an expert system or deep knowledge reasoner which could be developed along these lines. The knowledge base is built up in many layers with interactive connections within and between the layers. The user sees various types of facility available across the user-interface. One mode of operation is to answer questions about the structure or operation of the target system. Another function is performed by the builder module which enters information into the knowledge base to build up device models as the user enters structural and functional descriptions. The user can also control an envisioning module which will run the knowledge base in a simulation mode, thus producing predictive behaviours and behavioural scenarios for examination. The envisioner can also be run against a matching process to compare actual behaviour with the expected normal mode of behaviour. This gives diagnostic reasoning facilities. As an extension of this, the prognosis module could produce a report on what might happen if the system continued to run with faulty components or erroneous connections.

Two advantages gained from reasoning based on first principles are worth stressing. Most existing computer-based diagnostic systems assume relatively fixed structural relationships and deal only with first order deviations from specified functions. By contrast, a good engineer is able to consider different potential structural changes in the domain and discover corresponding changes of function. Such reasoning abilities will provide powerful diagnostic capabilities in future expert systems. The other advantage of a first principles approach is that the commonality of operating functions between many different devices can be expressed and exploited. Thus, it should be possible to build a mechanism library of universal primitive components and their functions, which can then be used to assemble more complex devices. This should offer a way of accumulating and transferring expertise across many related applications.

Although the techniques described in this section are also applicable in fields like medicine, telecommunications, software development and cognitive modelling, it seems likely that engineering, especially the mechanical and electronics branches, will be a particularly suitable field. Engineering seems a very promising domain for this kind of system because all man-made machinery is so well defined. In general, objects in the natural world are not easily specified in terms of good functional, structural or even geometric descriptions, whereas man-made artifacts, such as those found in engineering components and electronics, are extremely well-defined, and if they are machine-made they usually have a complete description. The very nature of engineering practice is helpful in providing structural and functional knowledge. This suggests that areas such as plant control, mechanical engineering systems, electronics, computers, and all sorts of machine-made devices are strong candidates for modelling with deep knowledge reasoning systems.

In addition, the range of potential applications is large, including: diagnostic and factory monitoring equipment, intelligent mechanism support for CAD, failure analysis, and planning systems. Qualitative simulation is likely to be useful in both analysis and synthesis roles. Diagnostic reasoning has been discussed but envisionments will be equally useful for exploring the behaviour of a new device design. Thus, planning and design systems should also benefit from first principles reasoning about device behaviour.

To summarize, we see that shallow knowledge represents short-cuts that efficiently combine evidence and make input/output associations, while deep models are able to support detailed analysis of system behaviour. Deep systems are also able to predict the consequences of events and take the reasoning one stage further. Shallow systems are much easier to build and get going but first principles reasoning is more general and therefore should be more valuable in its application.

Studies on deep knowledge for engineering devices will increase our understanding of mechanism modelling and representation. It will also add to the store of knowledge for designers and engineers who wish to implement computer-based tools for engineering.

8.7 Summary

1. Expert systems are a particular variety of knowledge-based systems. They are characterized by narrow but detailed expertise in an application domain and the ability to explain their reasoning. Most expert systems involve consulting sessions as their means of giving results.
2. There are two main classes of expert system problem — analysis and synthesis. These correspond to diagnosis and design respectively.
3. Generate-and-test is a useful constraint-based technique for both design and diagnosis reasoning systems.
4. Expert systems are now popular commercial products, but there are often problems with their performance. The shortcomings of many expert systems stimulate major areas of research in AI.
5. The creation of explicit knowledge is perhaps the most important benefit of expert system work.
6. There are many opportunities for expert systems and other forms of knowledge-based systems in all aspects of manufacturing and production engineering.
7. There will be important roles for opportunistic planning and execution monitoring systems as future manufacturing moves away from centralized control towards more distributed environments.
8. There are two classes of system operation for manufacturing applications: consulting systems for designers and managers, and automatic systems for factory-floor situations.
9. Deep knowledge methods deal with causality in systems and use common sense and first-principles reasoning.
10. Qualitative reasoning methods appear to offer many new facilities for modelling engineering devices and mechanisms.

8.8 Further reading material

Some of the issues raised in Section 8.2 are discussed in 'New research on expert systems', by B. G. Buchanan, in *Machine Intelligence* **10**, edited by J. E. Hayes, D. Michie and Y. H. Pao, (Ellis Horwood, 1982), pp. 269–299.

A tutorial article giving principles for matching different problems to different styles of expert system is 'The organization of expert systems: A tutorial', by M. Stefik *et al.*, *Artificial Intelligence* (1982), volume **18**, pp. 135–173.

Many interesting studies on deep knowledge topics can be found in the following two collections of influential papers:

(i) *Qualitative Reasoning about Physical Systems*, edited by D. G. Bobrow (North-Holland, 1984).
(ii) *Formal Theories of the Commonsense World*, edited by J. R. Hobbs and R. C. Moore (Ablex, New Jersey, 1985).

The papers by J. de Kleer and J. S. Brown, and B. C. Williams are relevant for device-centred models, while the work of K. Forbus concerns the development of process-based qualitative reasoning.

Chapter 9

Errors, failures and disasters

To conceive a plan and to carry it through are two different things.
G. Polya

Every week, in happy little family homes, something falls off, bends, buckles, breaks, dissolves, crumbles, collapses, bursts into flame or explodes.
C. Parker

9.1 The importance of automatic error diagnosis and recovery

In the scenario in Chapter 1, we described a future robot assembly cell that was capable of automatically building quite complex assemblies from a range of components. When the human supervisor asked for a report, the system was able to give both a diagnosis of any errors which had occurred and an account of any repair work that had been applied to rectify faulty assemblies. In robotic assembly any kind of failure, be it due to a faulty component or a workcell error, is a potentially serious problem and hence facilities for automatic recovery are extremely desirable. Such automatic methods should not only reduce the down-time (when maintenance staff work on the cell) but could also provide protection from serious damage during operation. In this chapter we will examine the problems of error recovery and review the requirements that are necessary for our scenario to become a reality.

It is useful to notice the different ways that errors can be defined. We do not use the word 'error' in the sense that is used in feedback-control systems, that is, a small *deviation* from a set-point or desired value. Feedback-control systems deal with errors that are defined as deviations from measured variables within an acceptable tolerance band. In such systems, adaptive error correction or compensation techniques are well understood and are frequently designed into a system as a feature of its normal operation. Instead, we use the term 'error' in the sense of an undesired event that is outside the system's normal operating range. Thus, our usage refers to all kinds of operating failures, including unexpected and novel events that have not been experienced before.

Notice that there are many situations, other than industrial automation, where the concept

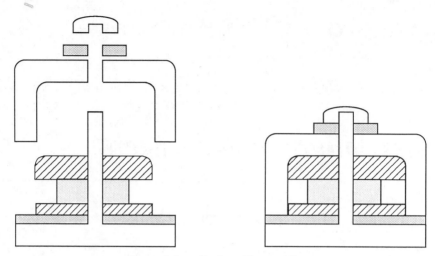

Figure 9.1   Real world assemblies

of automatic recovery is also very desirable and valuable. In undersea, space, nuclear and other hazardous environments, it may not be possible to call on full operator assistance and errors are bound to occur. Correcting unforeseen events will be an everyday task and autonomous methods will be essential. Consequently, error recovery is a very general topic of considerable importance. We will see that the concept embraces some of the most fundamental AI problems and future error recovery systems will require the use of a range of AI techniques. We suggest automatic error diagnosis and recovery is a classic case study for the use of AI techniques in industrial automation.

Many assembly tasks involve the stacking of components in the form of a concentric assembly. A typical operation will consist of a series of steps: acquiring a component from a feeder, transporting it to the assembly station, approaching along the axis of the assembly and mating with or without applied pressure. Figure 9.1 reminds us of the contrast between a typical industrial assembly job with an 'assembly' from an idealized AI world of the kind we have been using in our examples so far. Clearly, the industrial world is quite different from simplified textbook examples and we must be careful that we maintain realism and authenticity in our application of AI techniques. If we are going to deal with real-life applications then we must consider the differences in moving from the idealized world to real-world problems. There are many assumptions that are made in AI work and if the differences are so great that the known AI methods are no longer applicable then we could be in serious trouble. Unfortunately, it is usually not very easy to determine whether a chosen AI technique is or is not fully applicable to a particular problem. This is a key point, because if we dedicate a great deal of time and effort in developing a solution which is superficially successful, but at a deeper level is fundamentally inappropriate, we may well discover far too late that it is incapable of dealing with some new but vital aspects of the problem. Consequently, whenever possible, we should always try to assess the 'transfer value' of different theoretical methods for industrial application.

9.2 Classes of errors

There are three main classes of error that can be identified in a robot assembly cell. These are *hardware errors*, *software errors* and *operational errors*.

Hardware errors occur in all kinds of mechanical and electrical mechanisms, in control systems, in sensory devices and in electronic and computer systems. They are caused either by component failure or by design faults. There is a considerable literature on reliability technology that deals with the analysis of component failure rates and the design of reliable fault tolerant systems. The universal technique for reducing the effects of failures and faults is to introduce some form of redundancy. Redundancy methods are the basis of all forms of fault tolerant behaviour. This can range from redundancy at the lowest levels (e.g. component duplication), up to system redundancy (e.g. tandem operation of twin computer systems). At either extreme the basic properties of redundancy provide the essential characteristics needed for higher reliability performance.

Software errors occur through design faults in programs. Because programs are (theoretically!) fixed and time invariant, at least during the operation of a robot workcell, they will not exhibit intermittent or component failures, analogous to hardware faults. However, the principles of software engineering teach us that all programs contain design errors: bugs. The conventional wisdom goes — 'You can prove the existence of a bug, but you can't prove that a program is bug-free'. A common difficulty is that programs that have previously appeared to work can sometimes fail in almost identical situations. Methods to deal with this problem involve very careful control of the software production cycle, including testing, debugging, maintenance, and highly organized design methods and program-verifying techniques to reduce errors at the design stage. However, all of this effort is directed at fault *prevention*. There has been much less work on systems that can tolerate errors. In those systems that do tolerate software errors we find that redundancy is again used as the main technique. Redundant sections of code are used to reproduce the results of other modules and cross-check their answers.

Operational errors are the physical errors that occur in the robot task environment. These are not software or hardware errors but refer to a range of various faults in the configuration of the workcell. Common faults are defective components, alignment drift and jammed feeders. Most of these errors can be traced back to design problems. Redundancy cannot be considered as a viable technique here, especially at the higher system levels; notice that 'stand-by' robots, even if economically possible, could not access all of the operating space of a failed robot and therefore might well be prevented from fully reproducing the task. Such problems are often dealt with by a nearby machine operator or, in the long term, by re-design of the cell by production engineering staff. This is an area that is ripe for the development of automatic error recovery techniques.

The remainder of this chapter will concentrate on operational errors. As well as its importance for industrial manufacturing we believe this class of error will be predominant in the hazardous environments mentioned earlier. Notice that complete recovery from an error may not always be possible but we would like, at least, to diagnose the cause and provide some reasonable explanation as to why the failure occurred. Although we only consider operational errors, it is important to realize that the diagnosis methods can be extended to cover hardware or software errors, provided there is enough information available.

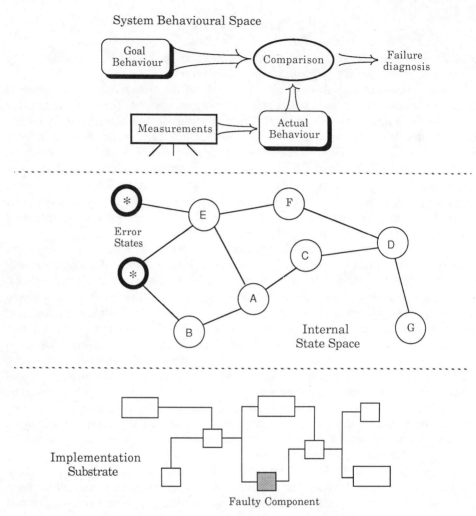

Figure 9.2   Failures, errors and faults

9.3 Observed behaviour and internal states

Let us now define the terms 'failure', 'error' and 'fault'. Figure 9.2 shows a symbolic representation of these ideas.

A *failure* is a difference between the actual behaviour of a system and the desired or expected behaviour. Thus, failures are measured by some form of operational or *behavioural* description. A *state* can be viewed as an internal configuration of a machine. An *error state* is, then, an internal configuration that is unacceptable as defined by the correct operation of the machine. Finally, a *fault* can be considered as a low-level failure of some sub-system. In other words, a fault causes a machine to get into an error state and the failure behaviour is a manifestation of the error state. Notice that a fault-tolerant system may have several internal faults but by avoiding getting into any of the error states it may

Errors, failures and disasters 179

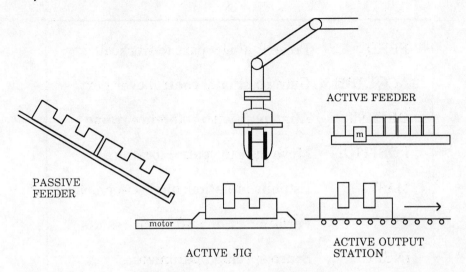

Figure 9.3 A robotic assembly cell

manage to perform up to specification with no failures being experienced. We can thus distinguish between realized error states and latent error states. This is why redundancy is so useful: when one component fails, error states can be by-passed by using another component.

Our terminology involves a kind of recursive definition as a fault in a system can also be viewed as a behavioural failure of a component at a lower level of cause and effect. To give a hardware example, a blown transistor may be seen as the *faulty component* in a system (it caused the system to realize an error state and fail) but, if viewed as an independent subsystem, it has *failed* as its behaviour was below standard due to some particular internal effect in that transistor (i.e. its internal fault mechanism). Despite this apparent complication, this terminology appears to be quite useful in defining system failures and their associated error events. During a diagnosis-reasoning process, it is possible to follow an analysis from observed failure data through error states to faults, and this recursive argument can then be repeated at each sub-system or component level until a suitable degree of causal analysis is reached.

9.4 Failures in the assembly world

Figure 9.3 shows the main components of a typical assembly cell. This represents a general model of a cell and contains features that will be found in most similar situations. The cell will assemble components according to a program of instructions that implements the given task specification. Figure 9.4 lists a series of standard activities that comprise the elements of any task. We argue that this list defines a general set of sub-tasks that will cover most, if not all, robot assembly actions.

We are able to classify errors in various ways. The kinds of error that might occur in an assembly cell can be grouped according to the kinds of entity that we find there, namely: devices and objects. These are listed in Figure 9.5 for each item in the workcell.

Now for any sub-task that the assembly cell might be trying to execute, we see that the

FEED	Present a new part to workcell
ACQUIRE	Gain positional control over part
ORIENT	Align part with reference frame
POSITION	Move part to spatial location
MATE	Establish relationship between parts
TEST	Perform action and sense results
INSPECT	Sense specified parameters

Figure 9.4   Assembly activities

- Component defects
- Feeder malfunction
- Manipulator errors
- Faulty gripper/tool action
- Jig or fixture errors
- External interference

Figure 9.5   Operational task errors

outcome of a task is either that it:

   occurs correctly as desired, *or*
   it doesn't occur at all, *or*
   it partially occurs, *or*
   too much activity occurs, *or*
   a completely wrong event occurs.

Given these five outcomes for an event as a generalization of all possibilities, we can apply this to each of the different devices in the workcell and build up an error classification for possible failure events. For example, if we consider a feeder; the feeder could either operate correctly and provide a new component, it might provide no component, it might overfeed

```
Grasp errors      - no part acquired
                  - several parts acquired
                  - faulty orientation of part
                              - location error
                              - gripper slip

Slippage          - lost part
                  - orientation disturbed

Release errors    - part not released
                  - orientation error
                  - displacement error
                  - ejection of part
```

Figure 9.6 Examples of gripper errors

```
Number          - increase      - breakages
                                - separation
                                - feeder error
                                - external item

                - decrease      - lost parts
                                - merged parts

Size & shape    - defects
                - damage
                - wrong item
                - faulty processing

Location        - small errors  - drift - alignment
                                - friction
                                - tool action
                                - wear

                - large errors  - gripper action
                                - external influences
                                - feeder errors
                                - obstructions
```

Figure 9.7 Examples of component errors

and produce two components, it might underfeed and only partially present a component, or it might misfeed and jam.

Similar arguments could be used to produce different results for each of the other items listed in Figure 9.5. For example, gripper errors could include the problems of grasp failure, slippage and release failure. These are listed in more detail in Figure 9.6.

Another useful example is the case of component errors. Figure 9.7 shows an error analysis tree for the different possibilities that might occur with workcell components. As component parts are continually entering the workcell, new examples are constantly being encountered and so the probability of faults depends upon the quality of those components. Quite a wide range of errors is possible with component parts. The number of components can increase or decrease unexpectedly due to breakages or other interactions, the size and shapes of the parts can be faulty, and their locations can have such variation that they are beyond the uncertainty limits tolerated by the system.

9.5 Coping with errors

Error recovery is the process of diagnosing the events that have led to a particular state, deducing which of the agents or objects in the world caused the events and deciding what should be done to restore the system to a more workable state.

One way of handling errors is to analyse the failure possibilities at *design time* when the workcell is being set up, and then build an error-handling routine for each likely case. This approach is sometimes called *error-recovery programming*, but as errors have been anticipated and analysed *before* the system runs it is really a form of *contingency programming*. Such pre-programmed methods are widely used in industrial robotics, and are suitable for cells where long production runs are to take place. However, for flexible small batch production these error-recovery techniques will not suffice.

There are a number of reasons why contingency programming will not suit future flexible manufacturing:

1 If we examine the code for a typical workcell we find that as much as 90% of the instructions are concerned with error handling — and it is hoped that most of this code will never be executed! Considering the amount of work involved, we may find this an unacceptable overhead if we have to produce such code for every small batch.
2 Another problem with contingency programming is the risk that some error cases will be overlooked by the programmer. It is entirely the programmer's responsibility to find all these cases and there are very few tools to assist in this task.
3 Another problem is that the methods used are very specific and so there is no carry-over into other applications. The error-recovery routines are not generic and do not provide for the abstraction and transfer of techniques into other applications.
4 Also, these are implicit methods of recovery. The reasons why particular methods and procedures were used is usually not contained in the code. We would prefer a more explicit approach whereby the reasons for different actions are available.
5 Finally, another reason why such contingency programming is not satisfactory is that some serious errors, such as hardware equipment failures, cannot be protected. This is because the programmer will not have access to enough information about the failure cases of different manufacturers' hardware.

Errors, failures and disasters

Considering all these factors we see that more automatic methods for dealing with errors are needed. In addition to overcoming the above problems, automatic methods will potentially be able to catch the unknown, underspecified error, including rare cases where something serious may go wrong. We believe that these automatic procedures must involve *run time* error processing systems so that they offer protection from all classes of error in order to prevent expensive damage or injury. Even if all error contingencies could be considered beforehand, which they cannot, flexible manufacturing will demand very rapid changes of task program and it is unlikely that the enormous design overheads of contingency programming could be supported.

In automatic-error recovery there are really three phases. These are:

1 *Monitoring*, which involves sensors in order to (a) confirm that everything is working correctly and (b) to detect errors.
2 *Diagnosis*, in which a series of symptoms is analysed in order to reason about the cause of a particular error state.
3 *Recovery*, where a recovery plan is formulated and executed in order to repair or correct the diagnosed errors.

We will now consider each of these three phases and look at their knowledge and processing requirements.

Monitoring is a task that obviously depends upon the quality and quantity of the sensors that are available. Chapter 4 discussed the problems of sensor data fusion and the difficulties of coordinating and controlling several different sensors. In the robot-assembly context we are concerned with the integration of sensing into the activities of the workcell so that the sensors may best observe and check that the desired processes are being carried out. This demands the careful selection and application of sensors to the task so that high-quality and relevant sensory data are provided for checking task functions. Each of the task activities listed in Figure 9.4 will require different sensory characteristics. The matching of sensors to tasks is dependent on the nature of the assembly cell equipment, the available sensors, and the nature of the critical task parameters. Given that suitable sensors have been selected and installed, the next problem is in activating and recruiting the sensors at the best possible time. We recall Figure 2.8 which showed a sensory regime where a set of sensors was turned on and off according to the needs of the task. Figure 9.8 shows a typical pick-and-place task with sensory points marked during the trajectory of the manipulator. As the manipulator moves into different critical regions so the sensors must be activated or de-activated according to the need for sensory knowledge at that point in the task. This kind of dynamic sensory regime is one of the main requirements for a good monitoring system. In order for the control system to be able to achieve this level of control over sensory activation, it must have knowledge of the expectations that can be associated with sensory data at different points on the task cycle. This knowledge can be provided in the form of tables of relevance and expectation values that can be used to decide which sensors to interrogate. In this way, sensory signatures can be attached to critical points in the task cycle. If the sensor values move out of the range of these signatures then the monitoring system will be triggered and cause an error to be signalled.

This style of task-based sensory regime has been implemented in the author's research group using a frame-based formalism. Figure 9.9 shows a diagram that illustrates the frame

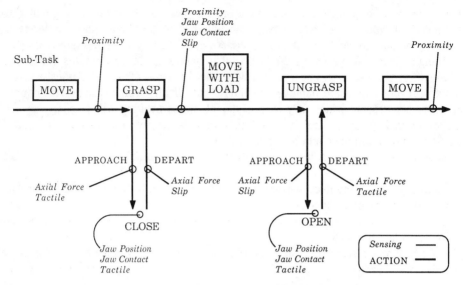

Figure 9.8 Sensing integrated into a pick and place task

structure where sensory sub-systems are automatically activated when particular actions are executed. Programming takes place at the task level by selecting different task frames to perform sequences of action that satisfy the required assembly job. The frame system has been designed to mirror the task structure of a common pick-and-place operation. The lower-level frames are automatically called when needed and integrate sensory monitoring into the task activity. The sensors are triggered by conditions within the frames and confirm the status of objects or the results of actions so that the next stage can be enabled. Frames are a natural knowledge representation for this problem as they can hold sensory expectations, they focus attention onto the current action, and they model stereotypical tasks with parameterized variations. The integration of action and sensing is the key feature that gives frame formalisms the ability to control dynamic sensory regimes.

Another approach to the problem of relating sensory knowledge to a task is to attach sensory information to the data about the objects in the environment. In this way, when a particular object is being handled or considered, its associated sensory frame can be activated to determine which sensors are to be used and to give settings and ranges for their relevant data.

The next stage of recovery is diagnosis and this involves the problem of reasoning about a series of sensory symptoms in order to propose the causes of those symptoms. There are two techniques which are relevant to our assembly world here. The method of generate-and-test, described previously, is particularly useful where all possible diagnostic senarios are to be considered. By using a generator, we may be able to build a series of hypotheses which represent each different case where the sensory signals may have arisen. By using all available constraints from the world and having knowledge of some possibilities that we can eliminate, we are able to reduce the number of hypotheses and hopefully obtain a small number for consideration.

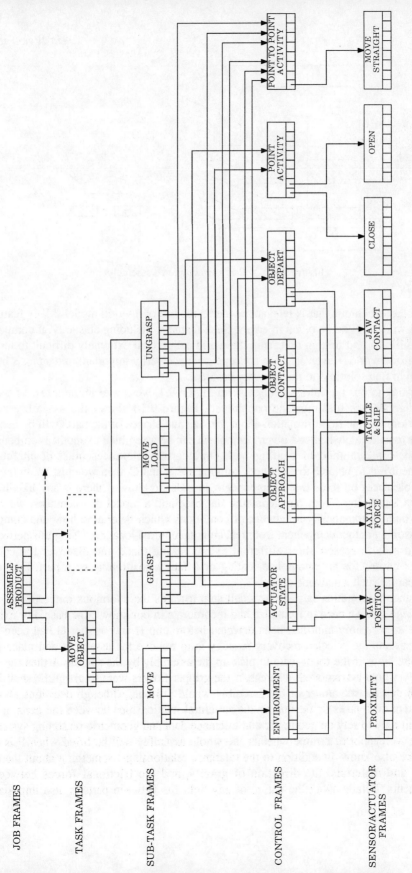

Figure 9.9 A frame based robot supervisor

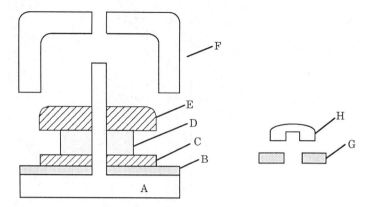

Figure 9.10   Partially completed assembly

The other technique that is relevant concerns the use of a world model. Many features of the real world will be involved in errors. Such features, including elasticity of components, greasy surfaces and foreign objects in the world cell, are all extremely difficult to model in the idealized world. However, these concepts are necessarily important in diagnosis because we wish to find solutions to the recovery problem.

Returning to our typical assembly task in Figure 9.1, we can examine a few of the real-world happenings that must be considered. Figure 9.10 shows the assembly nearing completion. Geometric information about the size and shapes of the parts will be necessary in order to calculate whether a given component can be assembled to another component. In addition, some method of dealing with engineering tolerances must be included. If component F is to be able to clear both components E and C, then we must know the range of the tolerances on these three components, in order to know if there is any likelihood of the parts interfering. From our geometric database and a model of tolerances, we will be able to build automatically a network of conditions which determine how one component can constrain another component from reaching its required location. Using this network we are then able to reason about different events taking place and discover the cause or potential causes for an incomplete mating or a faulty positioning of a part. Figure 9.11 shows part of such a network.

Tolerance information is just one small step towards the enormous range of real-world phenomena that we need to formalize and incorporate in our knowledge base. For example, suppose an assembly fault had been detected before clip H and washer G had been put in place. Assume the selected recovery strategy is to remove the assembly and place it in a reject bin. Now, if the robot were to pick up the assembly by the spindle A then the whole assembly would be removed, whereas, if the gripper fingers were to grasp the shell F, it is possible that the remainder of the assembly would fall out. Although decisions about the outcome of such tasks do depend on the structural relationships between the parts, it is not sufficient only to rely on geometric and tolerance data and geometric reasoning systems. In general, we cannot determine whether the whole assembly will be lifted when F is raised unless we also know, in addition to the tolerance relationships, something about the object masses and materials, the direction of gravity, and the frictional forces between the components. Details about the nature of any tight fits between parts is also important, as

Errors, failures and disasters

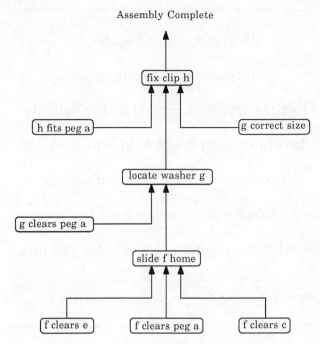

Figure 9.11 Part of causal graph for assembly

different elasticities will cause some parts to release easily while others may not.

By following these arguments and analysing similar case studies we soon reach the conclusion that comprehensive causal reasoning will be involved in any diagnosis system that is to provide powerful and complete diagnoses. The deep knowledge modelling methods of Chapter 8 are relevant here and we would expect such modelling techniques to be used in diagnosis systems that are able to handle real-world failures and detailed component faults.

The third stage of error recovery is the execution of actual recovery processes. The confidence with which we can perform an error recovery action will depend on the quality of the diagnostic information we receive. Clearly, a very complete and comprehensive diagnosis will allow an appropriate and effective recovery procedure to be selected and confidently performed.

If the workcell had been automatically programmed by high-level intelligent software of the kind discussed earlier, i.e. using a task-level language with 'what to do' rather than 'how to do' commands, we might expect any recovery planning to be undertaken by the same software. Error recovery then becomes just another task to be planned and executed. However, such synthesis of actions by automatic planning is an area fraught with difficulties. In Chapter 6 we saw some of the hazards of trying to apply theoretical AI solutions to industrial problems and we can summarize these planning difficulties again in Figure 9.12. Considering these problems, together with the fact that the diagnosis might not be perfect, it seems that quite different methods of recovery will be needed. The methods that are presently used tend to involve libraries of pre-compiled generalized recovery

> Multiple goals to achieve
>
> Multiple constraints to satisfy
>
> Multiple agents to control and coordinate
>
> Uncertainties in world & in world models
>
> Uncertainties in execution of actions
>
> Sensors give planning conflicts
>
> Recent events demand repeated replanning

Figure 9.12 Problems with the use of classic AI planners

> repeat test or inspection
>
> re-grasp part
>
> re-orient part
>
> repeat operation
>
> reverse operation
>
> reject part & feed a replacement
>
> recalibrate work cell fixture
>
> request replacement of faulty tool or jig
>
> reject current work & start new assembly
>
> reject batch & re-initialise work cell
>
> stop & issue alarm

Figure 9.13 Recovery actions

routines. The advantage of this approach is that the routines can be designed with relatively low risk, they can be tested independently beforehand and they provide the opportunity for easy extension and modification. Figure 9.13 shows a range of recovery actions, roughly in order of cost and complexity. These vary from minor corrections or repetitions of small movements through to full-scale rejection of work in progress and restarting the complete workcell. Although these recovery actions are similar to the actions which are programmed in contingency programming, the ways in which they are implemented are different. The recovery routines are kept as general as possible so that they may have wide application. In order to be confident that one of these actions should be performed, we will need to know much about the state of the workcell and the risk and costs of performing any action. This will require an expert system approach as described in Chapter 8. Following on from the diagnosis expert system, debugging and repair expert systems could respectively recommend suitable recovery actions and supervise their execution. Such systems will be able to take into account many variables, such as the severity of the error as diagnosed, the cost of attempting recovery, any risks of further complications, the preconditions to enable actions and any consequences that might follow on executing these actions. By these means, an expert system could assess the plausibility of a suitable recovery action. Information from the causal analysis produced by the diagnosis system would also be important for deducing whether some of the actions were reversible or not, or whether an object could actually be moved.

This style of system would give an effect which amounts to an opportunistic planning system. We would expect a blackboard style of architecture to be appropriate here, as the many variables that affect a recovery action would have to be weighed and considered against other actions in order to produce a recommendation with sufficient confidence. The blackboard model will allow quite different kinds of knowledge to be combined, and supports incremental planning.

The results of error-recovery actions could also be recorded so that learning mechanisms might be used to build up a history of the success or failure of different methods.

This kind of failure treatment is extremely effective in computing and other electronic systems where faulty modules are identified and simply removed and replaced. The difficulty with assembly and other manufacturing operations is that the error states have to be identified and corrected in the context of the working plant. It is not normally possible to remove a section of the plant and replace it with a 'spare'. Such 'board switching' strategies can only be done for individual parts or well-defined circumscribed sub-systems. The modular structure of electronic systems is a clear advantage here.

Another method of failure treatment, used in industry, is rectification, where components that are faulty are dismantled and re-built. This process is expensive but is economically viable where the components themselves have sufficiently high added value. In robot assembly it would be an extremely risky process to automatically disassemble much of the work piece — at least as risky as assembly! However, this may not be quite so necessary as with human assemblers because most errors will be noticed at the point where they occur if good monitoring and sensory regimes are used. Consequently, most errors will be damaged parts, missing parts, failure or wear of tools and assembly equipment, or failure of the hardware of the workcell. This last, being due to actuator or sensor failure, will most likely be beyond the system's recovery abilities, but a sufficiently powerful diagnostic system could at least prevent any further damage.

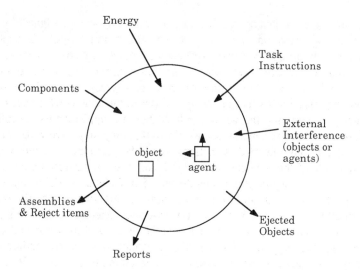

Figure 9.14   Abstract assembly cell

9.6 Building a world model

Now that we have seen the operational requirements of an error diagnosis and recovery system, we should look at the knowledge-base requirements of such systems. First, we will examine the available sources of knowledge and then try to identify the necessary characteristics of a knowledge base to support diagnosis and recovery.

When we consider existing assembly cells in terms of the knowledge which they have, we see that they really do not *know* what they are doing. The only 'knowledge' they have consists of (a) a list of variables defining a few locations in space and (b) procedures to execute movement instructions using those locations. Our aim is to give the system sufficient knowledge so that in some sense it 'knows' what it is trying to do and can reason about events when they are not carried out according to the expectations.

It will be useful to begin with the most fundamental and rather abstract concepts for world modelling. Figure 9.14 shows a representation of a system boundary for our assembly cell. The fundamental concepts are space, time, matter, energy and information. The diagram shows energy, information and objects entering and leaving across the system boundary. This is a view of the assembly cell from outside the boundary. Two of the transfers are due to error behaviour, that is external interference and objects being ejected out of the workcell. Within this world we would expect the laws of conservation of mass, momentum and energy to hold, and we wish to model events in the world with a suitable physics for assembly tasks.

Looking inside the system boundary we see various objects which can be classed according to their behaviour. Some of the objects are fixed and always remain in one spatial location while others are free and can be moved by the action of agents within the workcell. Some objects are capable of performing actions and so we also have active objects and passive objects. Figure 9.15 gives some examples of these different classes. Sensors are also objects but the transduction of sensory signals can be considered as an abstract activity

Errors, failures and disasters 191

	PASSIVE	ACTIVE
FIXED	FIXTURES WORK AREAS	FEEDERS CONVEYORS POWERED CLAMPS
MOVABLE	ASSEMBLIES & PARTS JIGS, TOOLS	MANIPULATORS END-EFFECTORS POWERED TOOLS

Figure 9.15 Classes of object

which takes place within a passive object. There are other agents within the world which are not attached to objects. These can be called 'influences': examples are gravity and magnetism. This type of model can be developed using the techniques of Chapter 8.

Each object in the world will have various properties. Some properties are intrinsic and do not change with the state of the object. Examples are material, colour and, usually, shape, mass, volume, etc. Other properties are extrinsic and consist of variables which may change with time. For example, an object's location, its relationship with other objects or the level of energy it contains, are all extrinsic variables which may be changed during the execution of an assembly task.

Our system model will have to record both the intrinsic properties of objects and their extrinsic variables. The collection of extrinsic variables defines the state of the system. Thus, there will be state variables for the actuators, for the environment, for the state of the assembly, for the different sensors and for all the component parts in different locations in the workcell. Most of the extrinsic variables can be entered at set-up time when the cell is being configured. The environment, the various actions that can be performed and the state and range of the sensing facilities will also be defined before execution time. The state variables which change during run time will be the locations of objects, the values received from sensors, the positions and forces applied by actuators and the energy levels that have been created within components.

Another way of looking at the assembly world is to distinguish the objects which can be sensed and those objects which can be controlled in some direct manner. Considering that some of the items are permanently within the workcell and some are transient, we obtain a diagram like Figure 9.16. This is intended to illustrate how some of the items in the workcell can be controlled, but there will always be others which we are not able to influence or change. Similarly, there will be some areas which sensors are able to monitor but there will also be others which are out of range of certain sensors. Considering that this will vary widely with each different assembly cell according to the design and manufacture of the cell, it is very difficult to avoid the arguments for building very specific software for error diagnosis and recovery. However, we see a generic approach as even more desirable because then the principles of recovery will be better defined and each application in each different area will be able to draw on these and tailor them according to its needs.

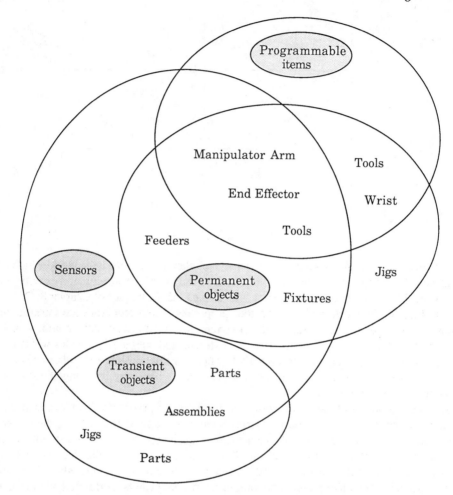

Figure 9.16 Components of the assembly world

If we now consider all the available sources of knowledge that we can find to help us in our knowledge-based system, we discover that there is quite a range of untapped material. Figure 9.17 illustrates different classes of knowledge that are associated with workcell equipment, product information and the assembly task. In each of these areas there exists extensive expertise which has been developed during the engineering activities undertaken to design and manufacture components and build workcells. When we look carefully at the assembly operation it is surprising how much knowledge is required for the achievement of such tasks. Much of this knowledge and information will be implicit knowledge and it is the purpose of our knowledge-based system to make this explicit for wider application and employment in other future knowledge-based reasoning processes.

Figure 9.18 illustrates the range of knowledge from different sources that can provide complementary and supportive information about the assembly process. Sensors give information on the immediate state of affairs in the workcell, while data on the characteristics of sensors give information about the likelihood of error or the confidence of

Errors, failures and disasters

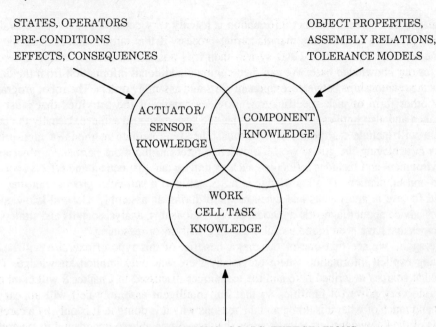

Figure 9.17 Assembly knowledge

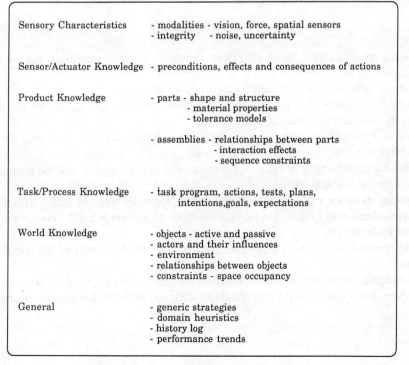

Figure 9.18 A robotics knowledge base

particular measurements. Product information is usually very detailed and complete as this has been produced to drive the manufacturing process. If this information is available as machine-readable data from a CAD system then it is possible to process it into different forms for our knowledge base. We can then compute additional information from the data, such as the relationships between component parts and assembly orders. The robot program, or any other form of task specification, provides details of the activities that must be undertaken and also implies the goal states that the system should achieve. Usually the task program will include test points and various other intermediate methods of measuring success in achieving the goals. World knowledge refers to the more permanent objects in the environment and includes influences such as gravity and magnetism and effects such as friction and lubrication. This knowledge is not specific to a particular product but may be invoked for use in analysis as and when required during an assembly. General knowledge includes advice about diagnostic methods, error relationships, analysis details and strategies and recipes that have been found useful in guiding analysis or reasoning.

Once again, we see that one of the major benefits of this type of exercise will be in producing explicit information where previously there was only implicit knowledge. The knowledge sources described here and the techniques discussed in Chapter 8 will combine to provide very powerful facilities so that our intelligent assembly cell will appear to understand much of what it is doing and the reasons why it is doing it. It is only by a careful analysis of the knowledge a system requires that we will be able to understand its potential abilities in processing information to satisfy given diagnosis or recovery tasks.

9.7 Summary

1 Error recovery involves three significant problem areas: monitoring, diagnosis and execution.
2 Automatic error recovery is a classic case study for AI.
3 Conventional contingency-programming techniques will not be satisfactory for flexible assembly with very small batch sizes.
4 Operational errors have many causes. There are different ways of classifying errors but what matters is that they can be modelled in the system's knowledge base.
5 Monitoring regimes must integrate the semantics of sensory signals into the meaning and purpose of the task actions.
6 Diagnosis systems will need to reason from first principles in order to deal with unexpected and novel failure events. The methods of Section 8.6 will become important for this in the future.
7 Recovery strategies should adapt general schemas to fit the details of the specific task domain.
8 All available knowledge sources should be recorded and employed in error recovery reasoning. Error diagnosis systems should be designed from this knowledge viewpoint.

Chapter 10

Better by design

Creative design is the central mission of the professional engineer
S. C. Florman

The previous chapter dealt with facilities which would enable an existing assembly system to improve its performance, particularly when errors occurred during assembly time. We focussed on the assembly scenario given earlier in order to discuss the knowledge requirements and processing abilities needed for such a scenario.

However, we have so far avoided an important issue. We have assumed the existence of suitable assembly cells but have not considered at all how they were designed, configured and built. Chapter 9 dealt only with the *execution time* aspects of assembly; we must also recognize the importance of *design time* decisions and processes. The design aspects of an assembly cell are very important as they will be responsible for much of its success when it is operating. Also, it is unlikely that all of the technology described in Chapter 9 would only be applied to the operation of existing cells, without being applied at the earlier stages when they are being created. In this chapter, we examine the issues of assembly cell design and configuration and consider how future systems might be prepared for operation.

10.1 A proposed assembly system

Let us consider that a product, consisting of several components, has been designed on a CAD system and the engineering staff are about to prepare an assembly cell for its manufacture. First of all, we notice that a 'part program' will have to be written for each component to define and control the machining and other processing operations involved. In an advanced system this would all be done within a computer framework rather than by being planned on paper. Consequently, we should expect the assembly sequence to be prepared in the same way. A robot assembly program could be created either straight after component design or even during the process by being integrated into the same CAD system. We assume here that the design system is either a very powerful corporate computer with access to all relevant product and process data, or, more likely, it is a powerful work station connected through a network to other systems which can supply the relevant data.

During the design stage all aspects of the product's creation and future life cycle should be considered. We would expect test data and sequences for test generation and measurements to be produced for mechanical and production engineering purposes. This might include test programs for computer-controlled coordinate measuring machines, instructions to quality inspection systems and sensor programs for supervision and monitoring. What we are describing here is a fully integrated design system which, once the product has been specified, is able to generate further control data for the manufacturing operations to produce that product.

In order to support such a powerful design system, there must be various knowledge-based or expert systems involved. During the product design stage such systems could advise on the cost of operations on components, efficient methods for assembly and design options that make manufacturing easier. The system could generate design parameters for human evaluation and preferences for style and efficiency. Of course, in order to produce robot programs, part programs or test sequences, such systems would have to know much about the robots, machine tools and test instruments that are involved. Information about these would be provided from library files. For example, a robot library would contain details of the robot systems, the fixturing arrangements, and various cell features which would be present in the particular assembly cell that the component is targeted for. This is already happening in the form of graphical robot simulators. These are available as commercial packages and can be used to examine the operation of a robot workcell on a visual display screen, taking information on selected robots from library data which have been previously prepared. These packages are used to help the designer visualize the layout and operation of the workcell and are useful for the detection of gross errors or inefficient sequences of operation.

Another feature that we would hope to see would be automation of the jigs and fixtures that hold the components during processing and assembly. This could involve either the use of special jigs which are designed at the same time as the components (a 'jig expert system' being used to design an appropriate jig for holding a newly designed component), or, alternatively, we might see a software clamping program being issued for a standard universal jig. In this way, we would expect to see the tooling of jigs and fixtures moving away from hardware design and machining problems towards soft fixturing. This will require libraries of programs which can drive variable jigs and fixtures to reconfigure themselves in different ways for clamping and holding functions. Such libraries of software and fixture data could also be coupled into design systems in the same way as for robot libraries and we could then design all of the significant cell features at the same time.

The next design time process to be automated would involve the selection of suitable sensors for the assembly task. This has been discussed in Chapter 2 and would involve an expert system that could select sensors for the critical areas of a task and configure them so as to provide the most appropriate data for monitoring assembly actions. Sensory expectations and relevancy data would also be calculated and prepared for down-loading to the run-time supervisor. The purpose of sensing is to reduce various uncertainties in the workcell and yet jigs and fixtures also reduce uncertainty in physical locations. So sensing and fixtures are complementary areas which can benefit from integration. When designing a fixture we may wish to consider whether a sensing technique could be incorporated to reduce the precision and complexity of the fixture, rather than controlling uncertainty purely by physical constraint. The trade-off between extra cost of sensor and simplified fixture could be evaluated by the designer, helped by the proposed knowledge-based system. It is likely that some aspects of design of the cell, e.g. for task execution and the selection of sensors, will always involve human guidance. But any support and automation of these processes would clearly be an advance over present methods.

It would seem essential that a simulator be built and incorporated at an early stage. The simulation of components on a screen is conventional; the extensions required here include physical assembly operations and processes. In this way, the arrangement of jigs, the design of the cell and the operation of the selected sensors could all be examined and checked. Notice here that sensors would need some care in simulation. Sensory data are not easy to appreciate by humans as such data do not have a physical realization in the same way that robots and objects do. However, it is equally important for simulation as sensory data are an integral feature of the task. This requires the simulator to have some means of displaying sensory information in a meaningful and task-related way.

Other work that the design system will do includes the planning of a suitable assembly sequence. This can be achieved by examining the relationships between the component parts and selecting an ordering which will be compatible with the operations available in the workcell. Various automatic planning systems have been proposed for assembly sequence planning and these could provide the basis for the necessary facilities.

Other information which must also be created includes the causal model for the objects in the world. This involves their structural details, their functional description and any physics or other operational data which may be utilized in design or diagnosis. The techniques described in Chapter 8 will be important here. The result of this will be a system that understands the manufacturing process and can assess the 'manufacturability' of the designed components (or the feasibility of a workcell configuration or robot task specification). Because the system links together so many knowledge sources it will be able to associate the shape of a part with the tools needed to process it and estimate the costs involved. Different databases will be examined to discover appropriate tools, processes and handling techniques. The reasons for different design decisions will be stored as designs develop, as will their costs and processing needs. In this way, the system not only holds the 'what' and the 'where' of a design, i.e. the structure and the form, but it also knows the 'how' and the 'why', i.e. its functions and the principles by which it works.

Temporal reasoning does not appear to be as big a problem for our system as spatial reasoning. It may be sufficient for the knowledge base to use time-stamped attributes in order to keep track of the temporal orderings of events.

The world model, as described earlier, will also have to be created and this will include the state space for the assembly world and all the intrinsic data associated with the entities in the world. We could expect much of the world model to be automatically generated by converting the available CAD data, both for the components and for the robot and assembly equipment. This would need to be in solid body form rather than based on line diagrams so that spatial occupancy could be represented and detected. This is important for planning paths between objects and for reasoning about geometrical relationships. Spatial constraints between objects, such as affixments, alignments, contacts and many other common engineering relations will also be deduced from such data sources and stored in the world model.

Tolerance representations will be needed for the components and other objects in the cell. Methods for manipulating tolerance models are required so that the accumulation of tolerance errors in a series of connected parts can be computed. This will be used to assess the likelihood of successful assembly during mating operations. In hard automation, component tolerances tend to be handled by relying on very precise positional locations. With flexible automation and the increasing use of sensors, positional uncertainties will be more easily accommodated and tolerance handling will thus move from the regime of hardware into software modelling techniques.

We notice that models of uncertainty will also be required for sensors and actuators because they are not perfect devices and have errors in their performance and operation. Because some of the activities of the workcell will depend on computed results from off-line design data, it will also be necessary to compute from the uncertainty models the likelihood of actions and sensory tasks being achieved. The knowledge-based system may also be able to supply some procedures for helping with the sensor fusion problem when different sensors are used in cooperation during a task. We would expect as much help as possible to be teased out of the design phase in order that the workcell could perform as effectively as possible. It is interesting that as these various stages become automated so the gap between design and execution tends to close. Although there is still a distinction between design time and assembly time, much of the information and knowledge which is used in these processes will be common to both. We see the whole emphasis shift from 'how to manufacture' questions towards 'what to manufacture' questions. As more of the production side of manufacturing becomes automated so there will be more computer-based knowledge available for better designs.

Figure 10.1 shows a diagram of the software involved in the proposed assembly system. The diagnosis and supervisory system are clearly components of the assembly cell but the planner and simulator could equally well be parts of a design system. We would expect this kind of integration to continue so that we have a community of expert systems which perform different tasks as information moves from the early design stages through to concrete instructions controlling manufacturing operations. The most important point is that all design files are available to any system. Such data may either be centralized or distributed but will be integrated into a coherent and comprehensive store of component knowledge.

It is also interesting that information does not flow simply from design to assembly, but experiences learned in the assembly process can be fed back to the design stage. Assemblies have to be verified before they are executed and the verification of a new cell would be performed on the simulator. The information gained from this activity will help in the design of further cells. Also, experience gained through diagnosis and error recovery will feed back through fault reports to further design stages. The 'design for assembly' approach to product design will now be directly supported and facilitated, as all relevant data on assembly, manufacturing processes and product requirements will be held on computer files. The process of designing components for ease of assembly is facilitated by this type of system as the coupling between knowledge for different purposes is much closer than in existing engineering practices. This integrated design environment should have a strong effect on future products.

We notice that learning techniques have not seemed very important. Although research on learning is a very active field in AI, the idea of learning in a production environment has

Better by design 199

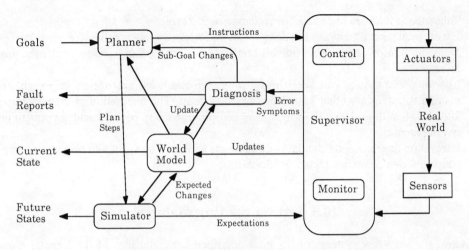

Figure 10.1 Proposed intelligent assembly system

little application. Small batches offer even less scope for learning (in the physical workcell) than does mass production. It seems AI learning methods will be more useful at a higher level, for trend analysis and quality control. In such management areas, learning will be useful for planning and strategy development.

The main theme of this chapter has been integration, and the two key factors which will facilitate integration, or could hinder it, are communications and standards. Efficient communications between different computers, different software systems, and different machine systems are vital in this type of approach to manufacturing. Various networks for factories and manufacturing plant are now available and developments in this area are essential if we are to enable knowledge and information to be explicitly processed at many different centres in the way we have described. The other vital factor is international standards. It is clear that no single manufacturer will make all of the software, computers or machine tools, and so, if we are to have a range of interesting opportunities for connecting different systems together, some form of standards will be necessary. This is not only true for communication protocols but standards are also vital for software packages, the interchange of data between systems, and the functional parameters and programming requirements of the machines themselves. Unfortunately, the computer industry does not have a history of adopting standards as readily as the other engineering professions.

10.2 Summary

1. Flexibility depends upon software development. The limiting factor for computer integrated manufacturing is not hardware or manufacturing technology but the availability of powerful software.
2. The integration of knowledge will enable designers and CAD systems to provide better assessments of the manufacturability of their designs.
3. Assembly cells will become more integrated. Tools and components will not be seen as different entities but all workcell elements will be subject to sensing and control.

4 Tolerance models are important for autonomous assembly.
5 Expert systems will support the designer in an integrated, cooperative framework. New areas are sensor selection, workcell configuration and programming, and execution monitoring.
6 Learning techniques and speech and language facilities are not as important for manufacturing as are other knowledge-based systems. The integration of manufacturing knowledge in the scheme described here will provide many benefits and support future developments.
7 Integration depends upon standards for products and interfaces and will be accelerated by cooperation in communications developments.

10.3 Further reading material

There exist very few systems of the kind described here. Because of the large size and expense of such projects, only the largest companies are able to afford the investment in research and development.

One interesting project in the UK is the 'Design to Product' project funded by the Alvey directorate and a consortium of companies. For details see 'Engineering design support systems', by R. Popplestone *et al.*, in the *1st International Conference on Applications of AI to Engineering Problems*, Southampton, April 1986. Another reference is 'The Alvey Large Scale Demonstrator Project, Design to Product', by T. Smithers, in the *3rd International Conference on Advanced Information Technology*, Zurich, 1985.

Some relevant research on assembly planning is described in 'Task planning and control synthesis for flexible assembly systems', by A. C. Sanderson and L. S. Homem-de-Mello, in *Machine Intelligence and Knowledge Engineering for Robotic Applications*, edited by A. K. C. Wong and A. Pugh (Springer-Verlag, 1987).

A project at the University of Rochester, USA, has investigated the problem of tolerances; see 'Towards a theory of geometric tolerancing' by A. A. G. Requicha, in the *International Journal of Robotics Research* (1983), volume 2, number 4.

Chapter 11

Towards a science of physical manipulation

*That's all the motorcycle is, a system of concepts worked out in steel.
There's no part in it, no shape in it, that is not out of someone's mind...*
R. M. Pirsig

From Chapter 1 to Chapter 10 we have come full circle. We started with a speculative scenario about advanced robot assembly and, after examining the many issues that such automation will entail, we arrived at a design for the software structure of a possible system. This chapter is an epilogue that reflects on the feasibility of robotics applications. We first review the different levels of motivation behind robotics research and discuss three research sub-cultures that form separate robot 'worlds': the mathematical world, the industrial world and the human world. Next, we consider a case study to illustrate the feasibility issue. Finally, we conclude with suggestions as to how the field might mature towards a more professional science of the control of manipulation in the physical world.

11.1 Introduction

There are many different kinds of machine tools, robots and factory processing equipment. The common thread running through all these is the idea of controlled physical change directed at a set of parts being manufactured. We have argued for task analysis as being a unifying approach that exposes the key issues and identifies common problems and methods between different applications.

The picture is often confused by the many different goals and methodologies that are found in robotics research and development projects. While some research is directed at 'frontier' problems, like automatic visual inspection, other groups are working on the industrial techniques needed to engineer new production systems. In addition, there are many forms of robot quite different from those found in the factory environment. These include underwater robots for oil-rig maintenance, systems for exploration and hazardous environments, space robots and mobile vehicles of all kinds. Robotics is a multi-faceted business which can be approached and interpreted from many different viewpoints.

Although all robot applications can be seen as different points on a spectrum of task complexity/difficulty it seems useful to identify three research sub-cultures that approach tasks so differently that they form separate robot 'worlds'. These are: the *mathematical world*, used in research laboratories for some of the more abstract investigations; the *industrial world*, as found in factories; and the *human world*, as experienced by humans in day-to-day life. Let us examine the characteristics of each world in turn.

11.2 The mathematical world

This area is characterized by the use of models; either abstract paper models used for systematic analysis or computer-based simulations that mimic the operation of a system. Experiments with models are a common method of investigating feasibility and performance when designing new equipment. The aim of modelling is to simplify the situation by removing non-essential elements and so discover the important parameters of the modelled system without expensive or dangerous full-scale trials. Prototype systems are usually built after sufficient experience of paper or computer models has been gained. Particularly in artificial intelligence work, robot models are often reduced to very primitive functional properties. This is because the goal of such work is to study the control problems associated with complex multi-dimensional tasks and consequently many engineering issues are largely ignored. Of course, simulations can be extremely authentic and appear to emulate exactly the real system, but all models can only provide a framework for examining parts of the full situation. The investigator or design engineer hopes that the most important aspects have been selected for the model, but we could argue that robotics models will always leave out some significant features. This is because robotics is so wide ranging in its physical and engineering implications that we will only discover all the problems when we start using a prototype in a realistic application setting.

11.3 The industrial world

In the factory, robots are used in well-defined roles and are engineered into their working environment. Measures are taken to reduce complexity by incorporating constraints into the task and the work space. This physical structuring of the environment simplifies the original task, but in a different way from the reductions used in modelling.

Let us consider a sample task: build a tower from three blocks. This task, in all its generality, we could call a *free task*. For example, in order to solve it we might need to know the properties of the blocks — their size, weight, strength, compliance, frictional features, etc. In the past, artificial intelligence researchers would use models that tend to ignore these 'engineering' factors and concentrate on programming the grasping operations and their logical relationships. Thus, the environmental features would be simplified and the task reduced by being made more abstract. We can call this a *formal task*.

On the other hand, an industrial version might use jigs and fixtures to locate and restrict the blocks, so that again the task is simplified. However, this is achieved by structuring and constraining the environment rather than reducing it to an abstract level. We can call this a *bounded task*. Figure 11.1 is a symbolic illustration of the differences of approach between

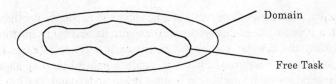
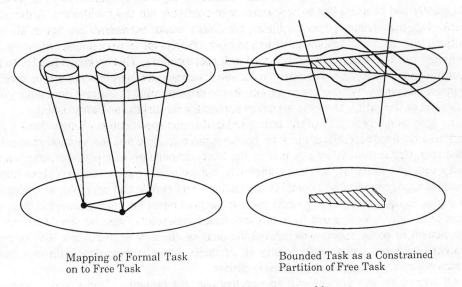

Mapping of Formal Task on to Free Task

Bounded Task as a Constrained Partition of Free Task

Figure 11.1 Mappings across worlds

formal and bounded tasks. Both are concerned with task reduction but by different means.

Recently, robotics research in AI has been moving towards the industrial area. Examples of this are seen in studies of grasping, collision detection and contact force analysis; all directed at real objects and robots. Some of the work on grasp planners incorporate frictional and inertial effects and has led to the design of new forms of dexterous robot fingers. On the industrial side there have been significant developments through the use of computer-based information systems. Industrial expertise is not often easy to capture and when a particular project has been completed, knowledge and know-how are often lost at the expense of future projects. This situation is improving with more and more use of on-line and computer-based records and documents. It is to be hoped that the top–down approach of the mathematical world (this includes most of AI) will continue to merge with the bottom–up development of the industrial world.

11.4 The human world

This domain covers all physical manipulation and mobility tasks that are normally carried out by humans, for example, the work of a car mechanic or nurse. This leads into the realms of science fiction as there really are no robots in existence for these tasks. However, this is such a popular image that we need to explore its feasibility for future research. The media, including the popular science press, regularly refer to robotics research as being on the brink of new breakthroughs that will produce general-purpose androids for public use. The motivation for such speculation is not hard to understand, but just how feasible *are* robotic devices in the real world? We already have robot bank clerks, tea-makers and car washers, so why is a robot nurse much more unlikely?

From our task-centred approach we can analyse the problem of human world robotics. Humans operate in a completely free task domain; at any time the parameters may change drastically and people adapt to new tasks or re-configure old tasks with ease. Although a bank clerk manipulates physical objects, the clerk's *actual movements* are not at all well defined from moment to moment. However, from a *functional* viewpoint many of the clerk's activities can be clearly defined in formal task specifications. This approach has led to the identification of the cash-dispensing role, and its consequent implementation in special-purpose machinery. We don't *need* a robot arm to dispense cash as the task does not require this level of flexibility. The same argument applies for tea-making and car-washing.

As soon as the generality of the task is reduced so the specification changes from a free task to a bounded task, the application becomes more feasible and the solution changes in character. Consequently, we can predict the likely automation of repetitious, mechanical tasks with low variability and low complexity, but we must be more cautious about highly variable situations involving complex interactions with people and operating in constantly changing environments. The 'general-purpose' android based on anthropomorphic design is seen to be quite unreasonable from this task-based viewpoint. Thus, we should beware of the market in hobby robots which leave the onus on the user to discover a use for them (rather like home computers). Nearly all of these 'home' robots deal with *free tasks*, because they are very general and loosely defined.

Of course, various devices will emerge that can, for example, clean a wide variety of floors, and other gadgets will be able to manage and control many home functions. But these robots will (a) *not* be anthropomorphic and (b) will be flexible only *within* their defined task range. They will have bounded tasks compared with the free task domain of an 'android'.

11.5 A case study

Consider the following requirement. A robot is to be designed for the home, which will fetch items on command. When it hears the command 'fetch aspirins' the robot is to bring the aspirin bottle from its place in the cupboard and hand it to its owner. This is a free task, because the robot is to be used in *anyone's* home. When we start to analyse the task we soon realize that the range of *significant* processes involved is considerable:

- Understand the command
- Plan navigation path to cupboard
- Execute motion to cupboard
- Open cupboard door
- Plan extraction operation

- Execute extraction and re-packing operation
- Close cupboard door
- Detect location of owner
- Plan and execute journey to owner
- Hand over the asprins.

Each one of these activities involves fundamental problems, most of which are the subject of major research efforts. Consider a few of the essential components of such a system:

(a) A large knowledge base containing information on the location of the cupboard, maps and plans of the house, arrangements and organization of objects within the house. All this is highly specific information that will change for different houses.
(b) A second knowledge source containing useful generic data, such as cupboard-opening mechanisms and their variants, i.e. procedures and methods for opening doors, sliding, swinging, lifting, pushing. This knowledge would be more widely applicable but would need to be complemented with specific variables.
(c) Powerful sensory-processing facilities. Sensory data would be continuously monitored and processed to guide navigation, locate and recognize objects, and provide feedback for correction of failed actions. This would have to deal with unconstrained random events and inconsistencies in the perceived environment.
(d) Natural language analysis abilities. In order to understand spoken commands extensive processing and powerful understanding systems are required. Major advances in speech processing and natural language understanding systems are now being made but this is still a very expensive facility.
(e) Planning capabilities. In order to locate the aspirins a logical plan of action has to be formed. If the aspirins are hidden, a suitable sequence of actions is needed to gain access to likely locations. Planning is further complicated by the fact that, in a free world, the options and conditions can change with time. It seems as if the rules of the game change while one is playing.

Notice that we have already limited the task by the emphasis on the home, i.e. the robot is restricted to house interiors. Just imagine the extra complexity if the robot was to travel outside. For example, consider navigation; imagine how the navigation problems change character in different scenarios — inside a house, in a street, in a factory or wandering around a country landscape!

In advanced robotics and AI work the many facets of the problem are usually factored out and tackled one at a time by individual projects and groups. Hence, many studies on planning, command languages, manipulation dynamics, vision processing etc. are reported in the literature and considerable strides are being taken. However, even when these areas can demonstrate success (and many of them can't yet do this at a commercial level) it is unreasonable to assume that their combination into one unified system is a simple next step. The integration of these areas is a major problem. Knowledge used in a vision project, for example, may be quite incompatible with that required for planning, and so the interaction of these sub-systems is yet another dimension to be investigated.

In the mathematical world we might take one aspect of the task and study it in abstract terms. For example, action planning could be intensively investigated to throw some light on the methods, resources, and processing required by the task. However, this form of problem reduction tends to keep us away from the real world by the emphasis on abstraction. Prototype studies will be eventually be necessary and will then force new considerations onto play.

The engineering approach to the task would be to analyse the fundamental requirement and provide this by any appropriate technology. A solution for our 'asprins' task might involve a microphone in every room, to detect simplified commands, and a delivery system (perhaps a pipe or chute) that can transport the desired item from an organized central store to the correct room. By concentrating on the task specification, using as many constraints as can be allowed, the final implementation of the solution may turn out to be quite unexpected, yet efficient and relatively simple. Removing or ignoring any anthropomorphic requirements drastically reduces the range of future tasks our robot can perform (i.e. it is less general), but renders the immediate problem much more tractable.

11.6 Feasibility and maturity

It is always important to assess the feasibility of any application before spending large quantities of time and money. We have argued that by refining the specification for any given task we can discern the nature of the problems and identify feasible solutions. By recording and reflecting on our experience, we can classify task areas into a framework that enables us to understand what are realistic goals for automation. In this way, we will learn how to balance ambitious objectives against attainable solutions.

There may be various reasons why a particular problem cannot effectively be solved by the application of computer technology. These include the following:

1 The domain may not be fully understood, e.g. economic or financial models are unable to accommodate all human influences.
2 The task may be too complex to compute in reasonable time, e.g. guaranteed solutions in chess involve the combinatorial explosion.
3 The available models may be inadequate, e.g. our incomplete understanding of human cognition limits our ability to design adequate models for psychology.
4 The sheer volume of relevant data may be too much to handle, e.g. in the human world, such as domestic environments, there may be so much information that access and processing may be severely hindered. Even the collection of such data is a daunting task.

The last factor is the one that particularly applies to the human world and often differentiates free tasks from bounded tasks.

This book has not pursued all the technical details of the methods described, but has tried to discern the fundamental nature of the problems. We have argued that flexible robotic machines that act on, and in, the real world require software control techniques that are different from conventional numeric and symbolic computational methods. The application of AI is clearly important to gain the capabilities that are necessary to support flexibility. Despite the fact that many of the problems are by no means solved, there emerges a general picture of knowledge-based manufacturing that offers great promise for future industries. Advances in this area will depend on continuing extensive research, especially in AI.

Towards a science of physical manipulation

As a general summary of the nature of robotic tasks, we finish with a list of key points:

1. Flexible systems must have powerful sensory capabilities. We cannot compute all the answers from our knowledge base, some information must be obtained directly from the environment.
2. Large amounts of domain knowledge are necessary. Complementary to the above point, we can not sense everything. Much knowledge is not even accessible to the sensory sphere.
3. Complete knowledge of the working environment cannot be achieved. By accepting this assertion we will be encouraged to build in methods that behave reliably with cases of missing information.
4. Only partial control over the working environment can be realistically achieved. There will always be some actions that we cannot perform or some obstacles to those that we can.
5. Actions and disturbances caused by external agents must be accepted as events that are to be accommodated. The real world is reactive, it is not a passive environment that we can manipulate at will.
6. Actions are not just the result of plans or computations; they have costs. Any action causes an effect on the world which might be damaging. Some actions will be irreversible, others cannot be repeated, while still others have wide influences on the world.

We hope that by recognizing these factors and by following this style of approach we may eventually develop a science of parts processing and manipulation in the physical world. The recent emergence of parts mating science is a beginning, but much more remains to be done.

Index

acquisition (of knowledge), 68–69, 150, 152
action, 7, 8
  influential, 123
  monitoring, 29
  recovery, 189
AI toolkits, 84, 154
artificial intelligence, 2–4
assembly cell design, 195–199
assembly, intelligent, 194
assembly monitoring, 196
assembly world, 191
associative nets, see representation – networks
attention focus, 23, 24

backing-up method, 110
belief systems, 96, 153
blackboard systems, 126–128, 139–141, 189
blocks world
  in vision, 61–63
  in planning, 112–118

CAD, 60, 147, 155, 173, 194-197
causal reasoning, 167, 171, 187
chaining, backward and forward, 77
CIM, 199
closed world assumption, 92
combinatorial explosion, 101, 110, 112, 206
communication, 9–10
  modes, 143–144
  natural language, 132,141,145
  systems, 199
contingency programming, 182
conversational systems, 8, 134-135, 142, 144

decision support systems, 147
deductive inference, 76–77
deep knowledge, 165–167, 171–173, 187
demons, 23
design, 155–156, 173, 195–198
design errors, 182
design-for-assembly, 198
diagnosis, 155,158,170-172,175-176,184, 198
distributed control, 88, 127, 139, 162

envisionments, 171–173
error, 7, 175
error recovery, 175–176, 187
  automatic, 183
errors
  assembly, 8
  component, 182
  gripper, 182
  hardware, 177
  novel events, 175
  operational, 177
  software, 177
execution monitoring, 158,162
expert systems, 148–162
  control, 158
  cooperating, 156
  debugging, 189
  design principles, 151
  explanation, 150, 152
  in electronics, 165–169
  in engineering, 173
  in mechanics, 169–172
  input/output, 160
  interpretation, 158

Index

prediction, 158
problems with, 152–154, 164-165
repair, 189
roles, 156–162
shells, 148
explicit knowledge, 151, 162, 194
extrinsic properties, 191

failure, 7, 178
failures in assembly, 179–182
fault, 178
fault tolerance, 177
feasibility, 1–2, 10, 204, 206
feature analysis, 31–34
feature space, 33
flexibility, 3, 9, 21
flexible machines, 5–7
frame system, 83–85, 183–185
functional knowledge, 171

game playing, 108–112
generate-and-test, 162, 184
graphical robot simulators, 196
graphs, 100
grippers, 14, 182
guarded moves, 24

Hearsay II, 138–140
heuristics, 67, 91, 92
Human Computer Interaction, 142–144

image constraints, 60–61, 65
image processing, 29–30
image understanding, 30, 53
implicit knowledge, 151, 192
industrial tasks, 4–5
inference
  Bayesian, 150, 153
inheritance, 80, 83–85
insertion
  peg-in-hole, 14, 18
inspection, 50, 196
  implicit, 57, 63
instantiation, 83
integrated design, 196–199
integrated knowledge, 169, 198
integrated manufacturing, 146, 195–199
integrated systems, 205
integrity (of knowledge), 91–94
intelligence, 3, 10
interpreter, 86
intrinsic properties, 191

jigs and fixtures, 196
just-in-time, 161

knowledge
  declarative, 71
  imperative, 71
  industrial domain, 89
  meta, 95

multi layered, 95, 150, 169, 171–172
  relative, 96, 135
knowledge bases, 67–97
knowledge engineering, 150
knowledge sources, 127, 139–140, 192–194

language processing, 132–135, 141
learning, 34,150,152,189, 198
LISP, 73, 82
local area networks, 146
local operators, 35–40

machine intelligence, 2
manufacturability, 197
manufacturing cells, 7
matching, 31, 57, 95,173
maturity, 2, 206–207
message structures, 143
model, 159
models
  causal, 197
  idealized, 99,176, 186
  structural, 167,171
  uncertainty, 198
  world, 186
moments, 44–45
monitoring, 155, 183–187

natural language, 9, 132

object classes, 191–192
object oriented programming, 118, 122
object properties, 29, 191
opportunistic planning, 161, 162, 189
opportunistic scheduling, 139

part programs, 158, 195
pattern recognition, 30–35
perception, 53–58
  in speech, 132–135
physical world, 6, 207
pick-and-place, 183
pixel, 27
planning, 8, 98, 128
  blocks world, 112–118
  frame axioms, 123
  goal directed, 118–123
  in manufacturing, 157–162
  protection mechanisms, 122
  rule based, 123–126
plausibility, 92, 189
predicate calculus, 73–75
problem understanding, 2, 10
process planning, 157, 158
production control, 158
production systems, 85–89, 149
  architecture, 85–89
  production rules, 88, 123–126
PROLOG, 73, 78, 82, 123
proprioceptors, 18

qualitative reasoning, 167, 171–173
qualitative values, 170
quality inspection, 196

real world tasks, 207
reasoning, 8
   assumption–based, 153
   common–sense, 167
   constraint–based, 164
   error recovery, 187–189
   first principles, 165, 167, 171
   industrial, 89
   non-monotonic, 94
   tentative, 153
recovery execution, 187
recovery strategy, 186
rectification, 8, 9, 189
redundancy, 179
relational analysis, 34
representation (of knowledge), 69–91
   frames, 83–85, 183–185
   logic, 72–78
   networks, 78–80
   procedures, 81–83
   production systems, 85–89
robot definitions, 5
robots
   first generation, 5, 7
   operating world, 6
rule based system, 149
run–time error processing, 183

scene analysis, 58–63
schemas, *see* representation – frames
search techniques
   beam, 105
   best first (A*), 106
   breadth first, 105
   depth first, 103, 114
   hill climbing, 103
searching, 99
   control, 102
   shortest path, 107
segmentation, 39–41
semantic nets, *see* representation – networks
sensing, 7–8, 13–26
sensor
   classification, 16–20
   configuration, 196
   coordination, 15, 21, 22–24
   expectations, 22, 183, 196
   function, 15
   integration, 20–21, 24, 184
   regimes, 22, 183, 184
   relevance, 22, 23, 183, 196
   selection, 13–20, 24, 196
   signatures, 22–24, 183

sensor control, 21, 22
sensor fault tolerance, 21
sensor fusion, 141, 198
sensor recruitment, 183
sensors
   external, 17
   imaging, 27
   internal, 17
   surface, 17
sensory constraints, 24–25
sensory flexibility, 15, 21, 25
sensory modalities, 17
sensory modes, 23–24
sensory overlap, 20–22
shallow knowledge, 165
simulation, 196, 197
slots, 83
small batches, 7, 9–10
spectrogram, 136
speech processing, 9, 135–140, 144–146
standards, 199
state space, 100, 162, 178, 197
structural knowledge, 171
supervisory systems, 198
syntactic analysis, 34

talkwriter, 134
task adaptation, 5
task analysis, 201
task-based approach, 2–4, 7, 204
task function, 18–29, 25
task reduction, 202–203
tasks
   bounded, 202, 204
   formal, 202
   free, 202, 204
   in sensing, 21
   level, 20
temporal knowledge, 81
theorem proving, 76
thinking, 7, 8
tolerances, 186, 198
transducer, 13
truth maintenance systems, 94, 153

viewpoints, 153
vision, 10
   computer, 27, 46–51
   human, 53–58, 64

Waltz filtering, 62
working memory, 166
world model, 186, 190–194, 197
   human, 202–204
   industrial, 202–203
   mathematical, 202